PARADISE LOST?

THE CLIMATE CRISIS AND THE HUMAN CONDITION

Paul Hoggett

PARADISE LOST? The Climate Crisis and the Human Condition
ISBN: 978-0-6488405-9-6
Copyright 2023 © Paul Hoggett
Published by the Simplicity Institute 2023
This text may be freely shared on a not-for-profit Creative Commons basis
Cover design by Andrew Doodson, Copyright © 2023

CONTENTS

Acknowledgements	v
Introduction	1
Section 1: Climate Change and the Modern Self	5
1. On Being 'Civilized'	7
2. A Creature at War with its Creatureliness	16
Excursion: The Ridge	24
3. Living in an Ideal World	26
4. Hyper-individualism	36
5. They Want All of You	44
Excursion: By the Banks of the M49	48
6. Time is the Enemy	51
7. Shame	56
8. Ressentiment	65
Section 2: Reactionary States of Mind as the Holocene Ends	71
9. Business as Usual	73
10. Fear of Truth	79
11. On The Inhuman	85
Excursion: The SS Armenian	94
12. The Armed Lifeboat	97
13. Reactionary States of Mind	103

Section 3: Facing Difficult Truths — 109

14. Feelings of Panic and Fantasies of Escape — 111

15. Grief, Mourning and Despair — 119

    Excursion: Scarp — 126

16. Staying with the Trouble — 130

17. The Tragic Position — 137

Section 4: Less is More, a New Ethic — 147

18. Paradoxes of Hope — 149

19. What is this Thing Called Love? — 155

20. On New Beginnings — 166

21. Revisiting Agency — 172

    Excursion: Bowerchalke — 181

22. Sustainable Activism — 184

23. What is More and What is Less? — 191

24. Revisiting Wellbeing — 201

25. Ordinary Ecstasy — 209

    Excursion: Meditations — 215

Acknowledgements

This book is the outcome of the discussions, workshops and talks that I've been involved in since the idea for what eventually became the Climate Psychology Alliance was put to me by Adrian Tait back in 2009. Besides Adrian, some others – Judith Anderson, Ro Randall, Tree Staunton, Sally Weintrobe – have been there from the beginning. I feel much gratitude towards all these companions and to many more I met on the way who directly or indirectly influenced my thinking. For more specific reasons I'm indebted to Tim Hockridge, Wendy Hollway, Breda Kingston, Julian Manley, Adrian Tait, Mary Travis and Rebecca Weston for valuable feedback on the first draft of the book. Finally, my thanks to Steve Thorpe and Sam Alexander for inspiration about the form the book has taken, particularly the use of colour imagery which is possible for relatively little cost in the e-book version.

Parts of the book draw on arguments that have appeared in some other publications. Chapter 2 draws on 'Imagining our way in the Anthropocene' *Organisational and Social Dynamics*, 2023, 23(1) 1-14. Chapter 7 draws on 'Shame and Performativity: Thoughts on the psychology of Neoliberalism', *Psychoanalysis, Culture and Society*, 2017, 22(4): 364-382. Chapter 8 draws on '*Ressentiment* and grievance', *British Journal of Psychotherapy*, 2018, 34(3): 393-407. Chapter 13 draws on 'Reactionary states of mind as the Holocene ends', *Psychoanalytic Inquiry*, 2023, 43(2): 119-129. Chapter 22 draws on 'Sustainable activism: Managing hope and despair in social movements', a piece written with Ro Randall in *Open Democracy*, 12 December, 2016. Finally, parts of Chapter 24 were originally published as 'Democracy, social relations and ecowelfare' in *Social Policy and Administration*, 2001, 35(5): 608-626.

INTRODUCTION

When someone comes to see you for therapy they are usually already embarked upon a long descent. Things are breaking down, old solutions (business as usual) are no longer working, the unintended consequences of their actions are cropping up all about them. But change is hard. Part of them will be stuck in denial, rationalisations will be offered, scapegoats will be sought. The past is so difficult to relinquish, the future is unfamiliar and uncertain. At first the promise of change seems to threaten disorder and chaos. Sometimes the old order is so strong that therapy results in a more painful stalemate and the relationship ends. More usually, therapy enables people to face difficult truths about themselves and others, and this makes them emotionally stronger.

Now, as the reader might have guessed, what I have just said using my 'therapist voice' might just as easily have been said by my 'climate activist voice'. For as the climate and ecological emergency deepens we are, all of us, embarked upon a long descent. Just how far we will descend, just how much destruction and chaos will ensue, none of us know. But of one thing we can be sure, our direction of travel is downwards, and the magnitude of the descent is in our own hands. This is the difficult truth we need to face. All else follows from this.

A lot of useful stuff has been written about the climate and ecological emergency by scientists, economists, philosophers and campaigners. Off and on I've been involved in politics since the 1970s and I've always felt that a psychological perspective can also deepen our engagement with such issues. Not the shallow, individualised psychology that is taught in most universities but a perspective which offers a deeper understanding of the modern human condition by drawing upon philosophy, literature, sociology, ecology and psychoanalysis. This is what this book sets out to do, this is what I think of as climate psychology. The catalyst for this book has been talks and papers I have given, workshops and conversations enjoyed since the idea for forming the Climate Psychology Alliance (CPA) began to emerge in 2009. The CPA (which now has members across the globe) offers psychological support to activists and concerned citizens, encourages therapists, clinical psychologists and other mental health professionals to work on climate issues in the public as well as professional sphere, and supports research and training in climate psychology.

The book comprises four sections and 24 short chapters. The first section asks what is it about Western civilization that has got Earth into this mess? It examines the founding myths of this particular civilization, specifically the unconscious idea of a paradise, an ideal world, which is just out of reach but might yet be found. Freed from need we became trapped by longing, a restless longing which has kept the wheels of progress turning whilst estranging us both from external nature, from our own mortal, creaturely nature, and from others. I examine how neoliberalism has exacerbated all of these trends leading to a kind of hyper-individualism which has had a profound impact on our psyches creating a 'modern self' which is oppressed by time, and which feels both special and worthless, privileged and resentful.

Section 2 examines the current stalemate between the old, fossil-fuelled capitalism and the new social forces. This is a period in which a variety of morbid symptoms can appear. The paralysis of liberal democracy in the face of the climate emergency is examined in terms of a culture of disavowal in which the truth can only be faced by splitting off our feelings, thus rendering the truth harmless and undisturbing. In contrast, far from being afraid of the truth, the new authoritarians either harbour nothing but contempt for it or cling desperately to old truths. Despite our cultural disavowal, signs of the worsening climate and ecological crisis continue to accumulate, generating a state of anxiety which feeds conspiratorial and increasingly psychotic thinking.

The last two sections of the book aim to provide resources to sustain an optimism of the will during the coming period, which may well be one of ecological austerity, social dislocation and political reaction. They examine the psychological resources (hope, love, courage, creativity, vitality) that will be needed in this period if we are to avoid a descent into inhumanity. They explain the importance of adopting a tragic perspective whilst simultaneously embracing the idea of 'natality', a love for life and new beginnings. They bring a psychoanalytic perspective to rethinking well-being in the context of a society where material growth is no longer possible let alone desirable and, in doing so, champion the idea of human capacities and the extraordinary nature of the ordinary.

I hope this is not an academic book. I've tried not to write in this way but it is difficult to shake off a particular way of writing which had been essential for success and recognition within this small but influential community. My endnotes, placed at the end of each chapter, are partly a concession to the academic that still lurks inside me but, perhaps more generously, I am sure they will help the reader who is interested to read more of the detail in my arguments than the main text allows for. But I've also introduced a set of 'Excursions' into this book which are mostly inspired by my walks. I love walking, especially when alone or with a dog. I find thoughts come to me, they inhabit me in a similar way to how a dream takes you over. So I ask you, the reader, to join me on these Excursions, to follow me without any irritable reaching for fact or reason into a different way of thinking. And finally, the reader will notice the book deploys imagery largely based upon photographs taken on my phone during these walks. As with the Excursions my hope is that these might facilitate a different kind of engagement with the book to what is convention.

SECTION ONE

CLIMATE CHANGE AND THE MODERN SELF

Each spring I wait anxiously for the return of swifts and swallows to my city. Harbingers of the great human migrations still to come, they find their way across the Sahara and then the Mediterranean and on to English shores. I remember one spring walking the Dorset cliffs near Golden Cap when a few early arrivals provided unexpected joy. To witness juvenile swifts excitedly screeching as they sweep above the rooftops of a Devon village, or to stand in amazement in a field as swallows career by a few feet away like fighter jets ducking below the radar, these are extraordinary experiences. To imagine a world without such creatures fills me with a sorrow beyond words.

Credit: Pixabay

How have we come to this pass? Our soils and oceans are becoming exhausted. Species are becoming extinct and biodiversity is being lost at an alarming rate. Countless marine and land-based ecosystems are becoming overwhelmed by climate change and other human impacts.

How have we become so estranged from nature? We seem to perceive it either as an infinitely giving host to be exploited in cavalier fashion or as a dump, a toilet to absorb and magically dispose of all our mess. But who is this 'we' that I find myself speaking of so easily? For the reality is that it is one particular civilization, one particular 'we', which has a specific responsibility for the crisis we now face. That civilization, developing in Europe for two millennia, is the Western civilization and it has Judeo-Christian roots. Unfortunately, despite its many achievements, it is this civilization which is also largely responsible for our estrangement from nature, from ourselves and from our fellow humans. It is this civilization which has also been responsible for the colonization, racism and class inequality which has defiled our world for several hundred years, and has brought us to the point where those in the Global South, who are least responsible for the calamity of climate change and ecological destruction, are those who will have to bear the worst of the consequences.

So when we talk about the impacts of civilization we must stop and think, which civilization, which 'we', is this and how do its citizens experience the world? What ways of thinking and feeling dominate this civilization? If what is sometimes unthinkingly referred to as the 'human self' is the product of this civilization what is the 'deep structure' of this self, one formed over the two millennia of development of Western civilization, and what have been the later impacts of modern capitalism and neoliberalism?

CHAPTER ONE

ON BEING 'CIVILIZED'

Within the climate movement it is often said that whilst climate change may not threaten the end of civilization it certainly threatens the end of civilization as we know it. So it is worth pausing to consider just what do we mean by civilization. Human civilizations are just a few thousand years old, a blink in 4.5 billion years of Earth's history. Approximately 11,000 years ago, when the last Ice Age was coming to an end, the Holocene period began. The uniquely benign climatic conditions inaugurated by the Holocene facilitated the development of stable human settlements and cultures across the earth for the first time.[1] This prompted the development of complex societies or civilizations, probably the first occurring in Mesopotamia about 3500 BC. There have been dozens of civilizations since then, across virtually all parts of the earth but one civilization, Western civilization, became prominent several hundred years ago and in the process destroyed or displaced nearly all others.[2]

The concentration of carbon emissions in the atmosphere during the Holocene period remained relatively constant at around 280 parts per million (ppm) until the rise of this new civilization.[3] The period of colonial expansion, the modernisation of agriculture and then the rise of industrialisation impacted upon emissions. These gradually increased until, after the Second World War, they began to grow exponentially (one of several aspects of what has become known as 'The Great Acceleration' in human activity). Today they are touching 420 ppm. The latest scientific research evidences nothing comparable during the last 2 million years, when temperatures were 5 degrees higher and alligators swam along the coast of Greenland. In only a few centuries Western civilization unlocked a massive reservoir of old carbon slumbering in the Earth and set about immolating it to power the modern world. As a result of the warming which has ensued, up to half of the tropical coral reefs on Earth have already died, 10 trillion tons of ice have melted, the ocean has grown 30 percent more acidic, and global temperatures are climbing steeply.

Nearly 15,000 scientists have now added their signatures to a warning with regards to the climate emergency. In an update, the original authors note, '(C)urrent policies are taking the planet to around 3 degrees Celsius warming by 2100, a temperature level that Earth has not experienced over the past 3 million years'.[4] The uniquely benign climatic conditions that made early civilizations possible are now being systematically destroyed.[5] We are now living in a new geological period, the Anthropocene[6], in which the conditions to support civilizations as we previously knew them no longer pertain. An unprecedented disturbance is afoot, and not just in the external climate.

The making of modernity

The period corresponding to the emergence of this new civilization, following the Renaissance in late-sixteenth-century Europe, is nowadays typically referred to as 'modernity'. It was powered by three interconnected developments.[7] First, the colonial and imperial conquest of most of the globe by a small handful of Western European countries. Second, the Enlightenment, with corresponding rapid advances in science and technology. Third, the destruction of 'the commons' (land, forests, lakes & rivers that everyone previously had access to) and its conversion into private property.

Whilst conquest initially led to simple plunder it soon made possible the acquisition of much greater wealth by harnessing and extracting resources from the conquered territories. Much of this extraction[8] required 'enclosure' of common land – jungle, savannah, mountain – which became controlled by giant new organisations such as the British East India Company. Meanwhile, advances in science and technology led to entirely new approaches to manufacture, transportation and agriculture, which required new forms of social organisation. Britain, the pioneer of these developments, was also one of the first European nations to practice enclosure of its own common lands, thereby 'modernising' agriculture and driving millions of now landless and destitute peasants into the new industrial cities, such as Manchester or into emigration abroad.

Before long the very idea of 'civilization' became appropriated by this new European elite, as if all that had preceded it was in some sort of way 'uncivilized' or even barbaric. To be civilized was to be modern. The violence of colonial conquest was justified as the necessary price of bringing 'civilization' to others.[9]

Conquest and civilization

The colonial process began with Portuguese and Spanish conquests in the Americas starting in the fifteenth century. British colonialism begins in earnest in the seventeenth century. By the middle of the following century Britain controlled over half of all the world's trade through organisations such as the British East India Company. Where soldiers ventured first, traders, businessmen and missionaries soon followed. The conquest and exploitation were justified in terms of a civilizing process - bringing civilization to peoples who were seen as less than human. By counterposing civilization to barbarism in this way, the barbarity inhering to modern Western civilization was obscured. Take Sigmund Freud and his classic book *Civilization and its Discontents*. You can read this text and have absolutely no inkling of any of the barbarity that made modern Christian civilization possible even though much of his book is about human aggression. In Freud's eyes, civilization embodied reason and he seemed to have little awareness that this civilization was built upon the extermination of indigenous peoples and the exploitation of premodern civilizations across the globe by colonial powers.[10]

Destroying the commons

As yet nobody 'owns' the atmosphere, nor the oceans, nor most of the great mountain ranges or the tropical jungles. The vast northern forests may 'belong' to specific countries such as Russia or Canada but not much of this is owned by private companies or individuals. This is what is meant by 'the commons', or more specifically, the natural commons. If it 'belongs' to anyone then it belongs to all of us. But this commons has been shrinking ever since modernity began – nature has been grabbed, enclosed and made private property. We see it happening before our eyes today as giant corporations such as Wilmar gobble up vast tracts of tropical jungle in Borneo to create palm oil plantations.[11]

In Britain the traces of this process are still discernible in the small tracts of common land often found outside towns and cities. Hundreds of years ago common land helped sustain the great bulk of the rural population but a gradual process of enclosure, which peaked in the period 1770 to 1830, led to the displacement of many people from the country to the misery of the slums of the growing cities. To give one example, by the middle of the nineteenth century in Manchester, average life expectancy was a mere twenty-five years.[12] Two centuries earlier average life expectancy in Britain was forty. And this was progress! The destruction of the commons accompanied the assertion of control over both the human and non-human. Both were becoming little more than a resource to be harnessed and exploited.

The triumph of reason

The modern era was built upon the split between science, on the one hand, and the humanities on the other. Freed from the humanities' questioning presence science could pursue the truth no matter what the cost was.[13]

During this period ideas of human exceptionalism (the human species has a special value and purpose), anthropocentrism (only humans have mind and soul) and the teleology of progress,[14] all become consolidated. Ethical questions concerning ends and purposes become eclipsed by a form of reason which was preoccupied with means, the 'why' of things became eclipsed by the 'how' of things. This is sometimes referred to as 'instrumental reason'.[15] As a consequence, once science and technology creates or discovers something new it becomes impossible or extremely difficult to 'undiscover' it. Thus the curse of the discovery of the atomic bomb and the development of nuclear weaponry – once the genie is out of the bottle it is so difficult to put it back. Because instrumental reason triumphs over substantive or ethical reason modernity lacks the power to hold back, to say 'let's not go down this route', 'let's leave this stone unturned'.[16]

The advances in science and technology that accompanied the Enlightenment led to the industrial revolution which was, at first, powered primarily by steam (the first commercial steam powered engine was Thomas Savery's water pump introduced in 1698). In other words, the extraction and use of fossil fuels, at first coal and later oil and gas, were the necessary conditions for powering capitalism. Capitalism has always been 'fossil capitalism'[17] – how hard it is to give up a habit of a lifetime.

Lawrence Weston

Probably the foremost exponent of the scientific method associated with the Enlightenment was Francis Bacon. His work has been subject to the most detailed critical scrutiny by the philosopher and historian Carolyn Merchant.[18] Integral to Bacon's approach was a mechanical view of nature.[19] The scientific method necessitated a violence to be enacted upon nature by the human mind, thereby fixing nature, dissecting and manipulating it. And in the process one part of nature, the human species, was privileged against all the rest of nature. This is 'human exceptionalism' in action. In the opening lines of her new book *How to Be Animal,* Melanie Challenger indicates where this has taken us:

> The world is now dominated by an animal that doesn't think it is an animal. And the future is being imagined by an animal that doesn't want to be an animal.[20]

God's chosen people?

It is worth considering whether the seeds of this modern way of thinking about our relation to nature might not be traced back two millennia to some of the Judeo-Christian founding myths underlying Western civilization. Christianity posits the existence of a single God rather than a plurality of Gods. But if there was one God then 'we', the believers, were the favourite children, special in the eyes of God. Thus the foundation of 'human exceptionalism'.[21] This was the idea that we were an exceptional species and the rest of the earth and all non-human species existed to serve 'our' purpose, which was ultimately God's purpose:

> And God said, Let us make man in our image, after our likeness: and let them have dominion over the fish of the sea, and over the fowl of the air, and over the cattle, and over all the earth, and over every creeping thing that creepeth upon the earth. *Genesis 1:26.*

Such statements provided part of the legitimation for humankind's dominion over nature. Nature becomes construed as separate and 'other', rather than something inextricably part of us, entangled with us. But if this tradition encouraged the belief that 'we' were an exceptional people, it also offered a warning to us not to imagine that we too could become God-like. The ancients understood that such hubris, the grandiosity underlying the passionate desire for mastery, led to 'the Fall'. In the myth of 'the Fall', Adam and Eve, ignoring God's warnings, eat from the Tree of Knowledge so that they can become like Gods themselves, and in doing so they are expelled from Paradise.[22] So 'the Fall' was construed as the 'original sin' and for millennia Christian religion called upon its imperfect and sinful subjects to know their place and acknowledge the love and wrath of the only perfect being,[23] God. An uneasy and often conflictual relation existed between science and organised religion until, as modernity advanced, the hold of religion began to weaken. Religion, that 'universal obsessional neurosis of humanity'[24] became displaced by reason. This was Nietzsche's celebrated Death of God:

> God is dead. God remains dead. And we have killed him. How shall we comfort ourselves, the murderers of all murderers? What was holiest and mightiest of all that the world has yet owned has bled to death under our knives: who will wipe this blood off us? What water is there for us to clean ourselves? What festivals of atonement, what sacred games shall we have to invent? Is not the greatness of this deed too great for us? Must we ourselves not become gods simply to appear worthy of it?[25]

How prophetic! Consider the following well known statement from Peter Thiel, founder of PayPal, ardent libertarian, transhumanist and Trump supporter:

> I remain committed to the faith of my teenage years: to authentic human freedom as a precondition for the highest good. I stand against confiscatory taxes, totalitarian collectives, and the ideology of the inevitability of the death of every individual. For all these reasons, I still call myself 'libertarian.'[26]

Peter Thiel is not alone. Ray Kurzweil, a key player at Google since 2012, joined Alcor Life Extension Foundation and was a great advocate of cryonics. It is rumoured that he is busy figuring out how to upload his brain into software before he shuffles off his mortal coil.[27] Do Thiel & Kurzweil believe that they are or can be supernatural beings or are they just afraid? The answer, in all likeliness, is both, for surely this is the paradox that we moderns have become, we are frightened hubrists. This is also the paradox of modern civilization – in its blind determination to master and control we can see the flight from vulnerability and dependence. And, unsurprisingly, this blindness is bringing about the very thing it feared, the return of the repressed – the reassertion of the forces of nature through fire, flood, storm and famine. This is what James Lovelock called The Revenge of Gaia.[28]

[1] Peter Frankopan has provided us with a comprehensive and illuminating analysis of the relationship between the natural environment (crucially including climate) and human history. I found his account of the emergence of early human civilizations in the period between 10,000 and 500 AD particularly fascinating. Frankopan, P. (2023) *The Earth Transformed: An Untold Story*. London: Bloomsbury.

[2] Among the most significant being the collapse of the Mayan, Aztec and Inca civilizations in the Americas following the arrival of Spanish and Portuguese colonists; the end of the Mughal Empire following the occupation and exploitation of the Indian subcontinent by The British East India Company in the eighteenth and nineteenth centuries; the collapse of the Qing Dynasty in China in the face of European, mostly British, imperialist expansion. Then, of course, there were countless examples of the extermination of indigenous hunter-gatherer and pastoralist peoples. Dan Gretton, for example, provides case studies of the German extermination of the Herero people in South West Africa and the British extermination of all Tasmanian aborigines at the end of the nineteenth century. Gretton, D. (2019) *I You We Them. Journeys Beyond Evil: The Desk Killers in History & Today*. London: William Heinemann.

[3] For a useful introduction to the history of climate change, particularly since the commencement of the Holocene period, see Lieberman, B. & Gordon, E. (2018) *Climate Change in Human History.* London: Bloomsbury Academic.

[4] Ripple, W., Wolf, C., Gregg, J., Levin, K., Rockstrom, J. Newsom, T., Betts, M. et.al. (2022) World Scientists' Warning of a Climate Emergency 2022, *BioScience*, 72(12): 1149-1155.

[5] IPCC (2023) AR6 Synthesis Report: Climate Change. Available at: https://www.ipcc.ch/report/sixth-assessment-report-cycle/ (accessed 18 September 2023).

[6] Although now widely accepted within the climate science community the concept of the Anthropocene still arouses considerable controversy, not least within the social sciences. Whilst Clive Hamilton offers a sympathetic exploration of the concept, Christophe Bonneuil and Jean-Baptiste Fressoz offer us a kaleidoscope of different perspectives. Hamilton, C. (2017) *Defiant Earth: The Fate of Humans in the Anthropocene.* Cambridge: Polity Press. Bonneuil, C. & Fressoz, J-P. (2017) *The Shock of the Anthropocene*. London: Verso.

[7] A very readable introduction to the emergence of modernity can be found in the first two chapters of Jason Hickel's recent book. Hickel, J. (2020) *Less is More: How Degrowth Will Save the World*. London: Heinemann.

[8] This extractive model of capitalism dominated relations between European powers and the Global South until the Second World War. From that point, radical anti-colonial governments took control and introduced policies to encourage public investment and protect domestic agriculture and industry. The neoliberal order soon got rid of this. The so-called Structural Adjustment Programmes tied loans and economic aid to the dismantling of such 'protectionism' in favour of 'free trade'.

[9] Probably the most comprehensive and uncompromising analysis of the violence 'intrinsic to Britain's civilizing mission' is provided by Caroline Elkins, Professor of History at Harvard University. Besides pioneering the use of the concentration camp (during the Boer War), this violence included 'electric shock, faecal and water torture; castration, forced hard labour; sodomy with broken bottles and vermin; forced marches through landmines; shin screwing; fingernail extraction; and public execution'. (p.28) As if to anticipate the reader's objection to this Elkins adds, '(F)ailure to confront these

practices diminishes the raw lived experiences in the empire and the legacy they left behind.' Moreover, all of this was perversely tied to a narrative of British exceptionalism, reform and progress (p.17). Elkins, C. (2022) *Legacy of Violence: A History of the British Empire*. London: The Bodley Head.

[10] Freud, S. (1929) *Civilization and its Discontents*. In J. Strachey (Ed.) The Standard Edition of the Complete Psychological Works of Sigmund Freud, Vol 21, 57-145. London: Hogarth Press.

[11] The Singapore company Wilmar are a major supplier of palm oil to the Chicago-based company Mendelez (formerly Kraft Foods) to use in products which range from Oreo biscuits to Cadbury's chocolate. In Borneo alone, 150,000 Orangutans have disappeared along with the tropical forest. https://www.nationalgeographic.com/animals/article/orangutan-habitat-loss-hunting-killing-borneo-spd (accessed 31 August 2023)

[12] See Hickel, *Less is More*, note 7 above, p. 50.

[13] According to the psychoanalyst W.R. Bion the pursuit of the truth in this single-minded way is the hallmark of arrogance. Bion, W.R. (1958) On Arrogance. *International Journal of Psychoanalysis*, 39: 144-146.

[14] A teleological view of history is one which sees history following inexorably in a pre-determined direction. One is bound to wonder whether modernity did actually replace religion or whether it simply continued the religious outlook but in a new secular form. Thus faith in progress slowly came to replace faith in God. Transhumanism and the belief that Artificial Intelligence will come to replace human intelligence could also be seen as a further manifestation of the teleology of progress, paradoxically one where humanity abolishes itself.

[15] Adorno, T. and Horkheimer, M. (1979) [1944]) *Dialectic of Enlightenment*. London: Verso.

[16] Naturally, all of these arguments are being played out once more with the rise of Artificial Intelligence and the Transhumanist movement.

[17] For an introduction to 'fossil capitalism' see Bonneuil and Fressoz, The Shock of the Anthropocene, note 7 above, pp. 199-206.

[18] Merchant, C. (1983) *The Death of Nature: Women, Ecology and the Scientific Revolution*. New York: Harper and Row.

[19] Using a historical and anti-colonial perspective Amitav Ghosh describes the struggle between this mechanistic philosophy and what he calls the vitalistic outlook of pre-modern society. Ghosh, A. (2021) *The Nutmeg's Curse: Parables for a Planet in Crisis*. London: John Murray.

[20] Challenger, M. (2021) *How to be Animal: A New History of What it Means to be Human*. Edinburgh: Canongate.

[21] Exceptions imagine themselves to be special and superior. Exceptions believe that they are exempt from the rules that apply to others, or that they are especially deserving or entitled in some way. The idea of the 'inner exception', something that lurks within all of we Moderns, is fundamental to psychoanalytic understandings of narcissism. Sally Weintrobe has recently explored the way in which this part of the self is given support and encouragement by neoliberal economic and cultural policies. Interestingly enough the idea of exceptionalism also figures prominently in political science. For example the concept of American Exceptionalism has been in use for many decades. Here a whole nation or people imagines itself to be exceptional in some way, and sometimes this is connected to the idea that 'we' have been specially

chosen by God to do his work on the earth. Weintrobe, S. (2021) *Psychological Roots of the Climate Crisis: Neoliberal Exceptionalism and the Culture of Uncare.* New York & London: Bloomsbury Academic. On American Exceptionalism see McCrisken, T. (2001) Exceptionalism. In DeConde, A., Burns, R. & Longevall, F. (Eds.) *Encyclopedia of American Foreign Policy, 2nd Ed.* New York: Scribner. On 'chosen peoples' see Smith, A. (2003) *Chosen Peoples: Sacred Sources of National Identity.* Oxford: Oxford University Press.

[22] Steiner, J. (2013) The Ideal and the Real in Klein and Milton: Some observations on reading Paradise Lost. *Psychoanalytic Quarterly*, 82 (4): 897-923.

[23] The many different ways in which Western religions grappled with notions of heavenly and human perfection were examined in depth by John Passmore. Passmore, J. (1970) *The Perfectibility of Man.* London: Duckworth.

[24] Freud, S. (1927) The Future of an Illusion. Standard Edition, 21, pp. 3-56.

[25] Nietzsche, F. (1882 (1974)) The Gay Science, trans Walter Kaufmann. New York: Vintage Books. Section 125.

[26] Thiel, P. (2009) The Education of a Libertarian. *Cato Unbound*, 13 April, 2009.

[27] Thiel and Kurzweil are examples of this movement which has become known as transhumanism. Transhumanists believe that the technology is almost available which will make bodily limits and constraints a thing of the past. I will provide a more extensive critique of transhumanism in my discussion of modern survivalist fantasies in chapter 14. Bill McKibben provides a good critical introduction to transhumanism in his recent book *Falter*. McKibben, B. (2019) *Falter: Has the Human Game Begun to Play Itself Out?* London: Wildfire.

[28] Lovelock, J. (2007) *The Revenge of Gaia.* London: Penguin. The misanthropic idea of nature taking its revenge on humanity recurs in Frankopan's concluding remarks (see note 1).

CHAPTER TWO

A CREATURE AT WAR WITH ITS CREATURELINESS

We are of nature and yet somehow also beyond nature or, if you like, we are a strange outgrowth of nature. One part of nature (the human species) has acquired the capacity to look upon the rest of nature and observe, study and intervene in it. It follows that this species has also acquired the capacity to be self-aware, to take itself as an object of contemplation and shape itself in a conscious way.[1] Let's be clear, it is not just we Moderns who have this capacity. Hunter-gatherers and pastoralists have this capacity too. But in their case it does not appear to break their connection to nature, nor their connection to the nature in themselves. In contrast, we Moderns seem to be creatures who are at war with our creatureliness.[2]

It is in acceptance of our mortality, frailty and fragility that we face up to our creaturely foundation. Every new scientific discovery, every extension of our control over nature and humanity, in short, every manifestation of our hubris is in part a flight from this unthought and unthinkable known. The reality is that we are born of nature and will soon return to nature. Our era (ie the era of the human) is but a gnat's blink in the life of this universe. Nature with all its limits and constraints is always closing upon us.

The active voice

Earlier I argued that the development of Western civilization corresponded to a progressive disconnection and alienation from nature as humans were encouraged to view themselves as an exceptional species in relation to a natural world which was there to serve their purpose. As modern society advanced and its capitalist form became entrenched, so nature became increasingly objectified and the way in which humans thought about themselves also gradually changed. Whereas in traditional societies human action was tightly bounded by constraint, accident and fate, under modernity people were encouraged to see themselves as active agents, as shapers of their own lives. Individualism strengthened this perception. By harnessing their own efforts and resources and by making judicious choices, modern individuals were increasingly held to be responsible for themselves, masters of their own fate. Failure was one's own responsibility, not the system's. If through what was seen as your own inadequacy, disability, infirmity or fecklessness you were unable to rise to this challenge then you were deemed to be undeserving of full respect and recognition, and in danger of being deprived of the rights attending to full human status.[3] Dependency became seen as the antithesis of agency.

In contrast the Modern was encouraged to believe that he[4] was in control of his journey through life, master of his own house. For this is an active, masculine voice, one that 'acts upon' the world and the people that comprise it. The impact of this modernist imagination can be experienced in psychotherapists' consulting rooms and the terror that many patients feel if they are not fully in control of every last, fine detail of their world. Their illusion is symptomatic of a culture in which the self is drawn repeatedly into a solipsistic retreat, alone against the universe, a kind of hermetically sealed 'I-ness' or 'we-ness'. Real agency requires engagement with reality, respect for the limits and constraints that reality poses, tolerance of frustration and hence perseverance and the capacity to acknowledge and learn from mistakes and failures. In such ways the will is strengthened. But in the ideal world, the world that we are led to believe we are entitled to, none of this applies. The capacity for real agency collapses and is replaced by what Sally Weintrobe calls the 'quick fix'.[5]

This identification of the modern self with the active rather than passive voice is deeply problematic when it gets mapped on to another opposition, ie subject/object. The human is then construed as if it were an entirely active subject whilst nature is construed as if it were simply the object for human interventions. In creating this dichotomy two vital dimensions of life are then repressed – nature as subject, and humanity as object.

Nature as subject

As a psychology student I remember having to study the early investigations of so-called 'animal psychologists' including the classics of Pavlov's dogs and Skinner's rats.[6] In those days these classics were held as exemplars of a scientific approach to psychology where variations in animal behaviour were closely monitored in relation to controlled regimes of reward and punishment administered by humans. The dogs and rats were quite unselfconsciously viewed as living machines – stimulus X produces response Y.

Perhaps it was no coincidence that running parallel to the evolution of modern psychology in the 1920s the factory farming of animals began with 'battery' methods of poultry farming. Immediately after the Second World War, at more or less exactly the same time as B.F. Skinner was publishing his major findings, legislation was passed in both the UK and the USA to promote more intensive methods of agriculture and thereby ensure 'food security'. By the 1960s factory farming of pigs, beef and dairy cattle was in full swing in these two countries. Factory farming exemplifies the process of objectifying nature which has crept right into the pores of the Modern self. What we have lost is a grasp of nature as a sentient subject, one that has feelings and agency in its own right.[7] There are several dimensions to this.

We imagine ourselves observing, analysing and sometimes (in the case of giving commands to domesticated animals) communicating with nature. We are oblivious to the many ways in which nature is busily engaged in sensing us, 'thinking' about us and communicating with us. For example, dogs, horses and other animals that are in close relation to humans can often sense when their human fellows are disturbed or unwell and can communicate what we call 'concern'.[8]

Nature speaks to us, but we have forgotten how to listen. I am not speaking here of nature as a bundle of our own projections, an anthropomorphised and sentimentalised nature. I am speaking of a nature which is no longer de-animated and objectified but is grasped in its living, vibrant, complex animism. We owe a debt to the BBC's *The Green Planet*, where the use of timelapse film technology brought the vitality of plants to life.

We imagine ourselves 'inside' a body and quite separate from the nonhuman. We are oblivious to the myriad ways in which the nonhuman affects us, invades us and infects us until a pandemic comes along and we wake up to what was always present. The reality is that bacteria swarm over every part of the body, through our orifices, into our gut. Via zoonosis pathogens leap from the animal to the human world, thus salmonella, Ebola, Aids, Avian Flu and now Covid, in fact over half of all pathogenetic human infections originate from the animal world. And now there is growing concern that zoonotic transmission is increasing as a result of the relentless pressure on the biosphere caused by human economic expansion.[9]

Inside our atomised imagination we observe a fish, bird or animal and speculate about what we think of as 'intelligence', as if the intelligence is located *within* the specimen under observation. In doing so we are oblivious to the distributed nature of intelligence, that is, to the reality that intelligence is supra-individual, located in the shoal, herd, pack or flock. In other words, intelligence resides in the system. Consider the dog as an example. Having ingratiated itself around the campfires of our ancestors and become a useful adjunct to sedentary farmers, it now often lives a life of domestic bliss, where it is pampered, warmed, exercised and fed, and usually doesn't have to do any work whatsoever in return.[10] In contrast, we have become so individualised that the human pack now often tends to lack any kind of intelligence whatsoever, particularly when it goes online.

Even when grasped in an atomised way, as an individual unit, many animals manifest what we Moderns think of as intelligence, including the capacity to adapt rapidly, to problem solve, use tools and symbolise. Indeed there is considerable evidence of the ways in which animals are adapting to the modern urban environment. Many of us in the UK are now aware of the way in which blackbirds and thrushes are shifting the pitch of their song in order to be heard above city traffic. But I did not know that thousands of white storks in Spain have stopped migrating to Africa because of the plentiful, all year food to be found in garbage dumps;[11] that house sparrows in New Zealand have mastered the automatic sliding doors of shopping malls and cafes;[12] and crows in Japan put walnuts in front of stationary cars at traffic lights in order to crack open the shell.[13]

Humanity as object

What also gets lost in the modernist imagination is the passive voice, where the human self exists primarily as the object of forces beyond its comprehension or control. We may be aware of the way in which powerful social forces, such as poverty or prejudice, impact upon us and yet be unable to do anything about it. Here the self faces real demands and real constraints which cannot be magically overcome but might be endured and occasionally resisted. But the experience of powerlessness can also impact upon a person's psychical integrity. The powerful other, whether a parent or partner or a dominant cultural group or class, invades the subject, gets inside their head, 'tells' them what they are thinking or feeling. Indeed the aggression needed to resist domination may, as in depression, become turned in upon the self. Depression is the most acute collapse of agency imaginable.[14]

As importantly there is an existential dimension to our object-like status. For the Italian philosopher Sebastiano Timpanaro, human frailty brought attention to the 'passive aspect' of relations between humans and nature.[15] To give an example, because we have bodies, we have pain. Pain can be occasional, unpredictable and intense, or like period pain it may be regular and predictable, and for some pain can be chronic and enduring. Whatever the shape and form, pain imposes itself upon us, it commands our attention, forces us to surrender to it. We may try to actively overcome it but so long as we have bodies, we are subject to pain.

We may spend much of our time attempting to control our bodies, manipulating them to serve our purposes through an active process of domination, but ultimately, they triumph over us. The question nature puts to each of us is this, can we surrender to her with grace and honour? Peta Bowden, an environmental campaigner and philosopher of care, put it beautifully:

> Illness presents us with explicit and indisputable evidence of the pervasiveness of chance and vulnerability as inherent structures of our lives. As the sufferers of assaults of happenstance, we experience the inescapable 'objectness' of our bodies.... The defenceless, thing-like fragility of body or mind is experienced in opposition to our purposes and values.[16]

When we speak of nature, of our alienation from nature, of our desire to master nature, of nature as this foe which must be conquered, we are normally speaking of nature 'out there', the nature of trees and seas. But what of the nature 'in here', the nature that resides within us, that is, the physical, bodily fabric of our being? Does not this second nature seem like an unnoticed intruder, an unwelcome guest who might, without a moment's notice, upset the peace? Are we not equally alienated from this nature, set upon a desire to conquer it and deny our actual enthrallment to it?

A couple of paragraphs earlier I said '(B)ecause we have bodies we have pain'. At the time of writing it I noticed that I hesitated, it occurred to me that the idea that we 'have' a body might itself be a misnomer and that it might be more accurate to say 'because we are bodies we have pain'. I wonder if this second way of speaking usefully puts me back in my place, stops me from getting too many ideas above my station. In contrast the first way of speaking makes it sound as if 'I' stood apart from and perhaps 'above' this body that I carry around with me, like some kind of excess baggage. Interestingly Bill McKibben notes that Robert Ettinger, the pioneer of cryogenics, 'found defecation so unpleasant that he wanted "alternative organs"'.[17] Being animal is such a drag, perhaps we can leave it behind. As we shall see in Chapter 14, some of the super-rich are getting ready to leave us, by being frozen in time until the crisis passes, hoping to upload their brains into powerful software, or by colonizing Mars.

Even as the object of misfortune and fate, perhaps especially then, we can find new strengths. Here the self is called upon to accept and endure rather than overcome, to seek equanimity rather than complain and protest about the shadow that nature casts upon us. Perhaps with the demise of traditional society and religion we Moderns have lost some of the arts of resignation and acceptance, of the capacity to go on, going on. To accept and value the unchanging, immutable, inflexible and incurable is deeply unfashionable in a culture which idealises cure, transformation and empowerment.

PARADISE LOST? The Climate Crisis and the Human Condition

We are then often the object of forces beyond our control or comprehension. As we shall see in Chapter 17, there is another area – the moral and ethical – where our agency is also greatly limited. Life presents us with insoluble dilemmas where there is no right thing to do. The Greeks knew of this thousands of years ago, it is captured so tragically in the figure of Antigone in the play by Sophocles.[18] And today, in modern society, the inherently conflictual nature of moral life lies even more exposed. Bonnie Honig, the feminist political and legal theorist, refers to this as the 'dilemmatic space'[19] of pluralist society where the different claims of race, class, gender, sexuality, disability, ethnicity and so on often rub up against each other, sometimes presenting predicaments for which there is no right answer. In such situations the best we can do is to fail better[20] – a far cry from the triumphalist rhetoric of much contemporary positive psychology.

---

[1] This contradictory quality of the human species – its capacity to be both in and beyond nature at the same time – has been wrestled with by many writers. Those I have found helpful include the following: Duguid, S. (2010) *Nature in Modernity: Servant, Citizen, Queen or Comrade*. New York: Peter Lang Press; Lasch, C. (1984) *The Minimal Self: Psychic Survival in Troubled Times*. New York: W.W. Norton & Co. particularly chapter 7; Kovel, J. (2002) *The Enemy of Nature: The End of Capitalism or the End of the World*. London: Zed Books.

[2] I first came across the notion of 'human creatures' when reading Donna Haraway's book *Staying with the Trouble* (2016). A few months later I came to realize that her reference to creatureliness was far from original. Ernest Becker uses the term in his book *The Denial of Death*, as does Nietzsche, who, in *Beyond Good and Evil* speaks of humans as both 'creatures and creators'. All three of these writers wrestle with the contradiction of the human condition in different ways. Haraway appears to 'solve' the contradiction by insisting that humans are just another 'varmint' (Haraway, p.169) along with microbes, plants, animals and even machines. This to me is a flight from a confrontation with what it means to be human, including some of the hideous things we are capable of. Nietzsche's 'resolution' is to fly off in the opposite direction: to be 'truly human' we must leave our animal nature behind. This is the accomplishment of the Overman or Ubermensch, the pursuit of perfection and immortality. As we shall see in Chapter 17, only Becker seems to be able stay with the contradiction. Haraway, D. (2016) *Staying with the Trouble*. Durham: Duke University Press; Nietzsche, F. (1966) *Beyond Good and Evil*, edited and translated by Walter Kaufmann. New York: Vintage Books; Becker, E. (1973) *The Denial of Death*. New York: The Free Press.

[3] There is a huge research literature cataloguing the way in which liberal democracies nevertheless denied many 'vulnerable' citizens their human rights, including the right to be protected from mistreatment and compulsory incarceration. Such citizens included those with learning and mental health difficulties, and often their mistreatment was compounded by sexist and racist stereotyping.

[4] My colleague Wendy Hollway has drawn attention to the gendered nature of the Modern self. Hollway, W. (2022) How the light gets in: Beyond psychology's Modern individual. In Hollway, W., Hoggett, P., Robertson, C. and Weintrobe, S. *Climate Psychology: A Matter of Life and Death*. Bicester: Phoenix Publishing House. This is a theme I return to in my exploration of agency in chapter 21.

[5] Weintrobe, S. (2021) *Psychological Roots of the Climate Crisis: Neoliberal Exceptionalism and the Culture of Uncare*. New York & London: Bloomsbury Academic.

[6] Ivan Pavlov investigated the salivatory response of dogs to food whilst undergoing electroshock. Lo and behold the dogs learnt to stop salivating! B.F. Skinner studied the way in which rats and pigeons learnt that a lever, if pressed, would either lead to a reward (food) or remove a punishment (electroshock). Together these animal experiments provided the foundation of what became known as 'behaviourism' – how animals and humans will learn to adjust their behaviour to pursue reward and avoid punishment.

[7] Philip Ball, the science writer who was an editor of *Nature* for many years summarises contemporary research evidence that perhaps even plants have sentience of sorts, and that many birds have their own 'theory of mind' and that even the honeybee uses forms of symbolic communication. Ball, P. (2022) *The Book of Minds: How to Understand Ourselves and Other Beings*. London: Picador.

[8] Sanford, E., Burt, E. and Meyers-Manor, J. (2018) Timmy's in the well: Empathy and prosocial helping in dogs. *Learning & Behaviour*, 46: 374-386.

[9] Editorial (2020) Zoonoses: Beyond the human-animal-environment interface. *The Lancet*, 4 July, 2020.

[10] The evolution of the relationship between human and canine is thoroughly examined in Bradshaw, J. (2011) *In Defence of Dogs.* London: Allen Lane.

[11] O'Mahoney, J. & Montero Sierra, D. (2023) In Spain storks' trash diet driven by climate change. https://phys.org/news/2022-07-storks-migrating-landfill-spain.html (accessed 30 August 2023).

[12] Ackerman, J. (2016) *The Genius of Birds*. London: Penguin Press.

[13] Kennell, J. (2015) How smart are crows? The answer may shock you. The Science Explorer, 11 December, 2015. http://thescienceexplorer.com/nature/how-smart-are-crows-answer-may-shock-you (accessed 30th August 2023).

[14] I discussed the ways in which poverty, class and prejudice destroy human agency in Hoggett, P. (2001) Agency, Rationality and Social Policy, *Journal of Social Policy*, 30 (1): 37-56.

[15] Timpanaro, S. (1970) *On Materialism*, trans Lawrence Garner. London: New Left Books.

[16] Bowden, P. (1997) *Caring: Gender-Sensitive Ethics*. London: Routledge. p.112.

[17] McKibben, B. (2019) *Falter: Has the Human Game Begun to Play Itself Out?* London: Wildfire.

[18] Sophocles (2009). *The Theban plays: Oedipus the king, Oedipus at Colonus, Antigone.* Fainlight, Ruth; Littman, Robert J. Baltimore: Johns Hopkins University Press.

[19] Honig, B. (1996) Difference, dilemmas and the politics of home. In S. Benhabib (ed.) *Democracy and Difference: Contesting the Boundaries of the Political*. Princeton, NJ: Princeton University Press. pp.257-277.

[20] Beckett, S. (1983) *Worstward Ho*. New York: Grove Press.

PARADISE LOST? The Climate Crisis and the Human Condition

Westward Ho! at low tide

EXCURSION

THE RIDGE

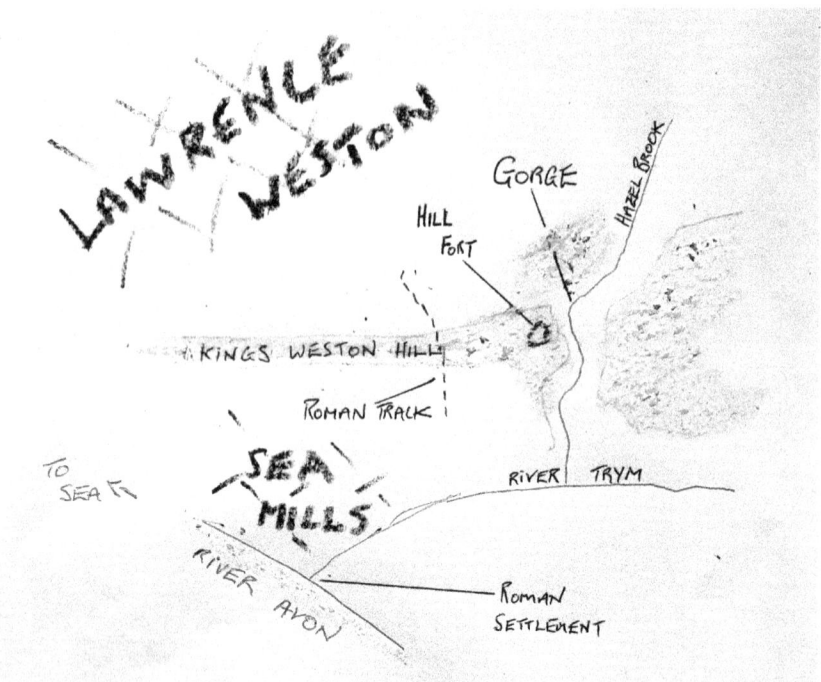

Walking up the hill from my house on the north-western edge of Bristol, the road comes to a dead end where a cobbled track takes you north up through the woods to what we locals call 'the ridge'. This track once connected Sea Mills (now a district of Bristol, originally the Roman settlement of Abonae) to Gloucester. After climbing about one hundred feet the track takes you out of the woods and onto the ridge, whose proper name is Kings Weston Hill. About a mile and a half to the south-west the ridge peters out where the river Avon flows into the Severn Estuary. A few hundred yards to the East the ridge plunges nearly two hundred feet into the gorge made by Hazel Brook, and then continues for another mile or so before being absorbed into a plateau which was once the site for the Bristol Filton Airfield. It is now home to a regional shopping centre and big new housing development. A few years back I discovered to my delight that the source of the Hazel Brook could be found in the car park of the Morrison's supermarket in the regional shopping centre. I never cease to wonder how such a small and short

stream (total length 4.5 miles) could cut such a massive gorge. After leaving the gorge it joins the river Trym (the title 'river' bestows upon the Trym a grandeur it does not deserve, it is actually smaller than Hazel Brook) and flows west for a mile until it reaches the Avon at Sea Mills.

Anyway, I've digressed, back to the ridge. On the very top of the ridge there's a kind of crossroads where the old Roman track traverses the ridge path. Turn east and slightly along the ridge path east towards the gorge and after about 60 yards you might just notice the path traverses a small undulation. In the summer when the grass is long you see nothing more, but in the winter you can see that this undulation forms a parabola stretching across the width of the ridge. This is the remains of an Iron Age stockade approximately 2300 years old. Walk on another 200 yards and you come to a much more pronounced earthen ditch and bank. This is the site of a fortified Iron Age farmstead which once occupied the prime location high above the gorge. Finally, half a mile back in the westerly direction along the ridge, remnants of neolithic tools have been found, including stone axe heads roughly 4500 years old.

Several thousand years of human history within ten minutes of my home and all on publicly accessible land. In fact Kings Weston Hill, the gorge, a community garden and much more are all part of the Blaise Estate, a huge public park in northwest Bristol. Here we have resident ravens, at least four ongoing badger sets, a magnificent 'beech cathedral' rising up the steep slope from the other side of the gorge, roe deer, buzzards and, every few years a visiting pair of sparrow hawks who bring carnage to the local avian population.

Being dog owners, my partner and I know just about every inch of Blaise. I reckon I've walked it about 2,500 times. And as I write this paean to it I can't help noticing that I am smiling to myself. So many experiences. Standing mesmerised on the ridge as the summer breeze blew waves through the long grass. Watching magpies taking on a marauding sparrow hawk. Gazing as seven buzzards formed a spiral above the valley. Listening to the cacophony of rooks as they settle down above the gorge at dusk. And that wind, that south-westerly which has whispered across this ridge for aeons.

All this wonder, all this time.

CHAPTER THREE

LIVING IN AN IDEAL WORLD

The nature of the Ideal

The shift from traditional to modern society led to the emergence of an increasingly individualised form of humanity, one which led to the formation of the modern self. This self was characterised by separation – separation from external nature, from its own creaturely nature and from other selves. Indeed, the very idea of 'the self' is to a great extent an outcome of modernity.[1] So when I refer to 'we Moderns' I am highlighting a particular form that part of humanity, particularly in the Global North, has taken over the last few hundred years. It is this self that is the subject and focus of psychoanalysis and modern psychology,[2] and not some hypothetical human universal.

As we have become freed from need, we Moderns seem to have become captured by longing. We seem hell bent on overcoming our earthly limitations. We shall become Gods. Perhaps Nietzsche was right, God is dead, we are supplanting him, paradise is within reach and this time there will be no fall from grace.

In my practice as a psychotherapist I have been struck by how tenaciously many of those I have worked with cling to the idea that there is some kind of ideal state of being in the world which has been lost or which is just out of reach. Sometimes this Ideal is imagined to have existed at some time in the past, for example, in childhood, particularly in a special relationship with a parent. Sometimes the Ideal is projected into the future, at some goal which, if achieved would bring not just satisfaction but eternal bliss. Sometimes this Ideal haunts the individual's present, particularly in terms of a preoccupation with time and the need to control it, as if time destroys the possibility of reaching the Ideal.

As we shall see in this chapter, the longing for the Ideal, the search for perfection, assumes so many forms in our lives besides those provided by material consumption and career success. Indeed, this longing has in no small way helped power the development of the productive forces that have generated modern society. Today contemporary capitalism has given credence, shape and form to this ideal world in a manner which no other type of society has come close to achieving. We have become a force of nature, thus the concept of the Anthropocene.

Henceforth I shall refer, wherever possible, to the Ideal as a noun with a capital 'I' to denote this phantom of the Christian and Modern imagination;

in other words the fantasy of paradise, a perfect place without limits or boundaries, one once thought of in terms of heaven, but now mostly thought of as a heaven-on-earth. I have given more detailed consideration, involving clinical examples, to the Ideal in a number of recent articles[3] where I came to see the value of distinguishing between two different relations to the Ideal that the self takes up – on the one hand, the fantasy of having it, on the other the fantasy of being it.

Having the Ideal

We Moderns are haunted by the feeling that we should have the Ideal. It can assume so many forms, from the ideal body to the ideal job, from the ideal home to the ideal wedding, and from the ideal family to the ideal death. Whatever the ideal it seems to promise a magical transformation in our lives. Sometimes these transformations are temporary, assuming the form of 'peak experiences' which can have an addictive quality.

    Eamonn was addicted to partying. The parties would typically start Friday evening and finish sometime Sunday afternoon. Usually fuelled by MDMA, Ketamine and electronic dance music (EDM), Eamonn could party for up to 48 hours without sleep. Perhaps more conventionally, Susan, in her late 50's, lived to travel. There was virtually no part of the world she hadn't flown to or cruised around. Each trip was preceded by months of anticipation but I noticed that after the trip, like Eamonn, there was a kind of temporary depression before the next destination arrived on her horizon. Each of these experiences provided a 'high' which, though transitory, was sufficient to quickly induce a longing for another similar experience. But in our imagination some kinds of transformation may be more permanent – the longed-for home in the country, the perfect job that had always been looked for, the religious/political community where one's 'true self' can finally be found. Some people spend their lives looking for the ideal partner, going through one relationship after another, never finding what they are looking for (because it doesn't exist). It is as if such people spend their lives searching, the seekers of the modern world.

    At the cultural level we are in the territory of the commodity object or experience, the allure of which pulls us towards it. Conventional wisdom states that human need creates an object in the mind which would satisfy that need – I am hungry, I think of food. But the commodity works in the opposite way, it is the presence of the object or its image that generates the need. We see it on a busy road, a beautiful white Tesla Model 3, or in a magazine we read about holidays in French Polynesia and look at the images of the beautiful beaches. And we feel the longing of unrequited desire. This is commodity fetishism, the search for the Ideal that keeps the system going.

At the political level we are familiar with the concept of 'ideals', ie the values that a person holds to, such as equality or responsibility, which offer guidance and meaning. It is only when it is adhered to rigidly, as if it were a fundamental truth, that 'an ideal' becomes '*the* Ideal'.[4] As we shall see in chapter 13, fundamentalists believe themselves to be in possession of the Ideal, a way of seeing which provides the total truth. This truth is often embodied in some kind of sacred text – the Old Testament, the Koran, the Talmud, *The Communist Manifesto*, *The Wealth of Nations*, *Mein Kampf* – which is understood as the literal truth.

Credit: Neom-venice-exhibition: Planned new linear city of Saudi Arabia

Being the Ideal

One should be the ideal parent, the perfect host, the selfless nurse or carer, the dynamic manager, the teacher whose lessons are never forgotten, the flawless lover and so on. All of these everyday manifestations of the Ideal are paralleled culturally in the rhetoric of achievement and success and in the celebrification process,[5] that fifteen minutes of fame parodied by Andy Warhol in 1968. Paradoxically the pursuit of fame can feel like a terrible burden. Two of the successful men that I have worked with were, from a very early age, 'the golden boy' of the family, destined to succeed like no others. They had to carry this burden of being their parents' Ideal throughout their lives, no matter what they achieved it never felt enough to themselves.

In neoliberal society success is itself idealised, to be a success is everything. Here is Ayn Rand, the literary inspiration of neoliberals everywhere, waxing lyrically about Dagny Taggart, the railroad executive heroine of *Atlas Shrugged*: 'her desire would never be satisfied, except by a being of equal greatness'.[6] Dagny finds her 'being of equal greatness' in Hank Reardon, a steel magnate. Together they become part of a small group of Ubermensch, who throw up their hands at the state of the world and wait for the coming economic and social collapse by holing out in Galt's Gulch, a libertarian settlement they create deep in Colorado. Just 60 years later and the world has indeed begun to resemble an Ubermensch paradise – neoliberalism having created a tiny elite of the super-rich[7] who 'enjoy' massive carbon intensive lifestyles[8] and are holed up in their gated communities, private islands and private statelets such as Monaco. The rest of us are 'the little people', yet little people who also might aspire to greatness.

The problem for the little people is that they are never good enough. As we shall see in Chapter 7, whilst psychoanalysts in the early twentieth century may have been in demand by guilt-ridden patients, today analysts, therapists and counsellors are besieged by shame-ridden individuals who feel themselves to be failures. Whilst the overriding feeling is of not being successful enough it is important to understand that success can assume myriad different forms. For some it is all about career advancement, of getting to the top and the imagined recognition that goes with this. For some it is about money, something which is seen as the measure of all worth and value. For others it is about power, the power of command, the pleasure of the flex.[9] These are all conventional models of success towards which the neoliberal subject is exhorted. But there exist myriad other everyday forms of success, similarly idealised, similarly impossible to achieve, which produce this feeling of not-enoughness. Among the many forms of not-enoughness I've encountered in my psychotherapy practice are people who have felt they were not caring enough, clever enough, tough enough, emotionally resilient enough, sexually successful enough, 'man' enough, 'womanly' enough, attractive enough or a good enough mother/lover/father.

Let's be clear, we all feel some of these things sometimes but I'm speaking here of people whose lives are dominated by these anxieties, anxieties that stop them from sleeping, can make them hide away and ruminate constantly. And they are gripped by these anxieties because the standard that they aspire to, the person they seek to become, only exists in fantasy, not in reality. This is the Ideal self, it lurks in all of us. One woman I tried to help saw her mission in life to make others happy. She had virtually no sense of her own need, that she might be entitled to some things for herself seemed sinful. She felt that she was a failure. I once asked her to imagine she was a high-jumper and to consider the yardstick against which she measured what success would look like for her. She replied that when she looked at this yardstick she realised there was no upper limit.

The point is that in an ideal world if you're not enough then you are nothing, or if not quite nothing then either something extremely small, insignificant and unnoticeable or something disgusting and shameful. In today's cruel parlance, you're a loser.

Dimensions of the ideal self

Given this separation, from nature, from one's own creaturely nature and from other selves, some psychoanalysts have wondered whether the origins of the modern self lie in a traumatic rupture.[10] We Moderns have been haunted by this place where there were no constraints or limits, where time and loss did not apply. We could not let go of the memory of this ideal world, and this longing for a perfect past pushed us ever forwards to discover more and more wondrous powers.[11] The gap between what is real and the Ideal has been both an affront to us Moderns and also a spur to achieve an ever more ideal world.

Omnipotence

The individual, group or society which imagines itself to be omnipotent sees itself as all powerful, of being able to overcome all limits, constraints or opponents. Very small children imagine themselves to have magic powers (and some much larger children too) but growing up involves learning to engage with reality. Engagement with reality is not the same as acceptance of or compliance with reality. Some limits and constraints may be overcome with sufficient knowledge, courage and tenacity. But the omnipotent strand within Modernity has led humanity to have such a striking impact upon the world that nature almost everywhere has been transformed into humanity's image – rivers straightened and dammed, oceans mined, jungles cleared, steppes converted to wheat and cereal production, the chemical composition of the atmosphere transformed – thus the adoption of the concept of the Anthropocene. Nations too may be gripped by omnipotent fantasies, hence the shock when, as with 9/11, these are disconfirmed by reality.[12]

Omniscience

Omniscience is the belief that one is all knowing. Again we can see a powerful strand of this in modern science and its periodic belief that it is close to solving the ultimate mysteries of the universe.[13] In premodern times this capacity to be all-knowing was a property of the religious 'adept' whose familiarity with the scriptures was such that they could see the ultimate truth beneath the messy surface of reality. In modern times, particularly since the development of the online world, adepts proliferate and have become the main carrier of conspiracy theories. As we shall see in chapter 13, Reed Berkowitz gives the term 'apophenia' to the tendency to perceive a connection or meaningful pattern between unrelated or random events. What follows is a kind of semantic promiscuity in which deeper and deeper meanings are 'seen' in otherwise pretty meaningless events. Thus, for example, The World Economic Forum presented a post-COVID economic recovery plan to the annual gathering of 'the great and the good' at Davos in 2020. Within months this innocuous piece of green tinged rhetoric sponsored by Prince (now King) Charles became the catalyst for a new global conspiracy called 'The Great Reset'.

Invulnerability

This involves the fantasy that one cannot be wounded, injured or harmed (physically or psychologically). It was manifest, for example, during Covid by many of those who, imagining they were invulnerable, concluded that they didn't need to wear a mask. Mask wearing was a restriction upon their liberties imposed by an arbitrary authority. In the process they completely overlooked the fact that they were vulnerable to Covid just like anyone else and therefore potential carriers from whom others needed protection. The fantasy of invulnerability makes us 'care-less' both towards others and towards the natural systems on which we depend.

Hubris

This often manifests as a dangerous overconfidence, combined with arrogance and excessive pride. It can lead to risk taking. One teacher I worked with, imagining himself to have outstanding talent, did some quite outrageous things with his classes. If he had been caught out, disaster would have followed. That he was never 'found out' simply added to his hubris. In a similar fashion we are taking outrageous risks with our atmosphere, despite the warnings about dangerous climate change exceeding 1.5°C.

Exceptionalism

Exceptions believe they are special because, they are the Ideal. Within the consulting room 'the exception' assumes many forms. The phantasy of being an immaculate conception captures the myth that one is a law unto oneself, dependent upon no-one, not even one's biological parents. Culturally it is manifest through an implicit assumption that one somehow is excepted from the rules and constraints that apply to all others. As the American real estate heiress Fiona Helmsley once famously put it, only the little people pay taxes[14]. As we saw in chapter 1, beyond the individual, exceptionalism also operates at the level of the group, particularly through the idea that one's group is somehow an exception, an imagined community which has a special place in history or has been 'chosen' in the eyes of God. Exceptions see climate chaos coming towards them and believe that somehow it will not apply to them, either because they belong to a people chosen by God and will therefore be saved or, more cynically, because they are rich enough to avoid its worse consequences.

Entitlement

Here the Ideal says to us, 'you should have me, you are entitled to me'. This sense of entitlement has two different sources. On the one hand, it is almost like a seductive and collusive voice which says 'yes, you deserve more than these others, you are special'. On the other hand, it can connect to the part of the self which feels that it is never successful enough no matter how hard it tries. Here the advertiser's phrase 'because you're worth it' appeals to the frazzled and exhausted self, offering reassurance, as if excessive consumption is a form of therapy which enhances wellbeing. So whilst unhealthy entitlement is often linked to greed, here this greed is powered by a sense of privation.[15]

At a cultural level the entitlement of Western citizens is linked to their blindness about the origins of their prosperity compared to much of the rest of the world and the way in which this prosperity has relied upon sustaining relations of economic and cultural dependency between nations of the Global South and the North. We know that we are in debt to the Global South, but when at successive United Nations Climate Change Conferences (COPs) it comes to putting our money where our mouth is, we somehow just can't seem to get our act together. Likewise, as individuals go to a restaurant, we hop on a plane, we refurbish our home, we update our devices, without giving a second thought to the uniquely privileged nature of each of these practices.

The two fundamental laws

I have written about the exception and the entitled as if they were somehow 'other' but they are to be found lurking, in varying degrees, in all of us Moderns. At a scarcely conscious level we imagine that if we really were the Ideal, or really had it, then we would be in paradise. In paradise there is no lack, no longing, no sense of time, no sense of loss. The grip of the Ideal lies in the force and power of the desire it provokes and the more it fails or frustrates us or seems beyond us the more we reach for it. Failure of the Ideal, what Janine Chasseguet-Smirgel referred to as the 'malady of the ideal', is inevitable. Failure to be the Ideal means that one is small, insignificant, inadequate or invisible; either one is all or one is nothing, shame and humiliation beckons. In contrast, as we shall see in chapter 8, failure to have the Ideal to which one feels rightfully entitled arouses a powerful sense of being wronged, of grievance and victimhood (victims are always in the right). The belief that the longed-for object has been taken, stolen or spoiled by another, substitutes for being able to acknowledge the loss of the illusion and mourn its passing.

It follows that we Moderns appear to have two primary psychological operating systems. According to the Law of All or Nothing, if I cannot be the Ideal (all) then I am nothing, and feel shame; according to the Law of Right or Wrong, if I cannot have the Ideal then I have been wronged, a victim of injustice, and this arouses a resentful and angry righteousness in me.[16] These processes are even more pronounced when we consider the way in which the Ideal operates at the societal level. As I mentioned earlier, fundamentalist religious and political groups imagine themselves to be in possession of the Ideal – a set of sacred texts or a political programme. They believe themselves to be wholly in the right, bearers of a set of fundamental truths, and unbelievers are wholly in the wrong – those who do not share in their Ideal are the damned and corrupted, literally superfluous, either to be abused and illtreated or to be disposed of.[17]

---

[1] Seigel, J. (2012) *The Idea of the Self: Thought and Experience in Europe Since the Seventeenth Century.* Cambridge: Cambridge University Press.

[2] Hollway, W., Hoggett, P., Robertson, C. and Weintrobe, S. *Climate Psychology: A Matter of Life and Death.* Bicester: Phoenix Publishing House.

[3] Hoggett, P. (2020) The malady of the ideal and the assault on nature. *Psychoanalysis, Culture and Society,* 25(1): 83-99; Hoggett, P. (2020) The grip of the ideal, *British Journal of Psychotherapy,* 36 (3): 415-429.

[4] The theme of the ideal has been present in psychoanalysis since Freud introduced the idea of the 'ego ideal', only to lose sight of it again as he subsumed it within the super-ego. Good analyses of the vagaries of the ideal in Freud's thinking have been provided by the following: Sandler, J., Holder, A. and Meers, D. (1963) The ego ideal and the ideal self. *Psychoanalytic Study of the Child*, 18: 139-158; Lansky, M. (1999) Shame and the idea of a central affect. *Psychoanalytic Inquiry*, 19: 347-361. Although psychoanalysis continued to pay considerable attention to processes of idealization whereby the self attributes ideal qualities to the other, the significance of the Ideal (as a noun) only re-emerged within psychoanalysis in the 1970s and 80s through figures such as Heinz Kohut, Janine-Chasseguet-Smirgel and Charles Hanly. Kohut, H. (1971) *The Analysis of the Self*. New York: International Universities Press; Chasseguet-Smirgel, J. (1985) *The Ego Ideal: A Psychoanalytic Essay on the Malady of the Ideal*. London: Free Association Books; Hanly, C. (1984) Ego ideal and ideal ego. *International Journal of Psychoanalysis* 65(3): 253-262.

[5] Turner, G. (2004) *Understanding Celebrity*. London: Sage.

[6] Rand, A. (1957) *Atlas Shrugged*. New York: Random House. p.220.

[7] A recent report by Credit Suisse estimates that 45% of the world's wealth is owned by the top 1% and within this 1% there are 215,030 ultra high net worth individuals with assets over 50 million dollars each, 55% of whom come from the USA. See https://www.credit-suisse.com/uk/en/articles/media-releases/global-wealth-report-2021-202106.html (accessed 30 August 2023).

[8] An Oxfam research report found that the world's richest 1% need to reduce their carbon emissions by 97% to comply with the 1.5 degrees target set by the Paris Agreement in 2015. See https://www.oxfam.org/en/press-releases/carbon-emissions-richest-1-set-be-30-times-15degc-limit-2030 (accessed 30 August 2023)

[9] A phrase used by the playwright David Hare to describe the motivation behind George Bush's invasion of Iraq. Hare, D. (2003) Don't Look for a Reason. *Guardian*, 12 April 2003.

[10] Wurmser, L. (2015) Mortal wound, shame, and tragic search: Reflections on tragic experience and tragic conflicts in history, literature and psychotherapy. *Psychoanalytic Inquiry*, 35 (1): 13-39.

[11] Becker connects this to the hero myth. The first section of his book is called 'The depth psychology of heroism'. Becker, E. (1973) *The Denial of Death*. New York: The Free Press.

[12] Clarke, S. and Hoggett, P. (2004) The Empire of fear: The American political psyche and the culture of paranoia, *Psychodynamic Practice*, 10 (1): 89-106

[13] James Lovelock gently mocked this kind of scientific hubris. Lovelock, J. (2007) *The Revenge of Gaia*. London: Penguin. pp.48-9.

[14] Fallows, J. (2016) Only the Little People Pay Taxes, The Atlantic, September 29th, 2016. https://www.theatlantic.com/politics/archive/2016/09/only-the-little-people-pay-taxes/623116/ (accessed 30 August 2023).

[15] These two different sources of deprivation leading to entitlement are examined in Gerard, J. (2002) A sense of entitlement: Vicissitudes of working with 'special' patients. *British Journal Psychotherapy*, 19 (2): 173-188.

[16] My hypothesis is that these two laws of psychic functioning are some of the principal manifestations of the psychological splitting process first outlined in detail by the psychoanalyst Melanie Klein.

[17] The connection between totalitarian thought and human superfluousness has a running thread in Hannah Arendt's analysis. Arendt, H. (2017 [1951]) *The Origins of Totalitarianism*. London: Penguin.

CHAPTER FOUR

HYPER-INDIVIDUALISM

Neoliberal policy

Neoliberalism is[1] an ideology which has dominated the thinking of business and government for four decades, starting in 1980. It asserts that by maximizing market freedoms the optimal conditions for the growth of capitalism are ensured. It reifies 'the market', turns it into a phantasm which like an invisible hand, if left to its own devices without government interference, would inevitably lead to the best use of resources, to the most efficient and vigorous kind of economy and society. Unfettered competition was the fuel, deregulated markets were the lubricant. Markets were to be freed primarily through deregulation and privatisation. As markets extended their reach, our ways of relating to each other and to nature increasingly became infected by the market relationship.[2]

Government was generally regarded with profound suspicion. Government officials, including front line professionals involved in welfare provision, were seen as essentially parasitical actors whose primary concern was to further their own interests,[3] constantly on the lookout for ways to waste taxpayers hard-earned money. Taxation beyond what was required to support the security of the state and private property was regarded as abhorrent, a drain upon the vitality of the entrepreneurs on whose efforts the health of economy and society depended.

For neoliberalism there was no such thing as the common or public good. The only good was the good of the individual and his family. This outlook was captured in Margaret Thatcher's infamous statement:

> I think we have gone through a period when too many children and people have been given to understand 'I have a problem, it is the Government's job to cope with it!' or 'I have a problem, I will go and get a grant to cope with it!' 'I am homeless, the Government must house me!' and so they are casting their problems on society and who is society? There is no such thing! There are individual men and women and there are families and no government can do anything except through people and people look to themselves first. It's our duty to look after ourselves and then, also to look after our neighbour.[4]

PARADISE LOST? The Climate Crisis and the Human Condition   37

Lawrence Weston

Neoliberal rationality

Neoliberalism was not just a set of policies. It was also a rationality,[5] that is, a way of thinking, feeling and being in the world. This began to alter the way in which we thought about and acted towards all aspects of life, such as how we should bring up our children, how we thought about nature and how we thought about and related to ourselves. In other words, as a rationality, neoliberalism began to seep into the pores of our culture, our everyday lives and our psyches. Increasingly we approached everything with an economic calculus in mind as if we were accountants busily measuring, enumerating and calculating inputs and outputs, costs and benefits. Suddenly the care of your old mum or dad became subject to such regimes. Think of what is involved in the care of an isolated, housebound old man with mobility and continence problems. Now compare it with the following:

(P)urchasers...would initially assess each customer and prescribe: how many visits per week, the exact time of day for each visit, exactly what tasks should be done on each visit, and the precise length of each visit. The prescription of fixed visit lengths (by which payment to a provider is calculated) sometimes leads to the label 'time-centred working'.[6]

And in such ways quantification destroyed what was once relational and human.

Nature is also subject to this rationality, including our approach to preserving it. Nature becomes seen as a resource, that is, as an asset for human use, providing 'ecosystem services', not as something with intrinsic value. And like any other asset it can be saved or banked. I visit a website on what's called 'mitigation banking'. It tells me that 'depletion of resources renders them a commodity' making them 'critical natural capital' and 'increasing their conservation value'.[7] So nature becomes increasingly subject to commodification, a process reflected in the spread of environmental economics programmes in the university sector.[8]

Ethics

Neoliberalism promotes an ethic of individual responsibility, of hard work, thrift and achievement. It is not that it is necessarily uncaring, it simply insists that the proper place for care and for human dependency is the family. Outside of the family neoliberal citizens are exhorted to be independent, to stand on their own two feet and to do this they sometimes need to 'toughen up' ('tough love' as the neoliberal technocrat Tony Blair put it). Moral behaviour is perceived as something which applies entirely to this personal sphere and, as we shall see in chapter 9, this reinforces the splitting of public from private life, which has given so much license to corporate irresponsibility and destructiveness.

Wendy Brown may be right to suspect that this emphasis on family values provides cover for patriarchy and the continued domination of white males,[9] but I wonder whether there may not be a contradiction within neoliberal conservativism here because in its pure form this ethic of individual achievement is thoroughly meritocratic.[10] 'The best' should succeed irrespective of extraneous factors such as one's gender or the colour of one's skin.[11] As neoliberalism became entwined with globalisation, capitalism – at least it's self-image – was becoming increasingly meritocratic, multicultural and ungendered. Prejudice was a barrier to business in a global corporate environment, a relic of the past.[12] It is quite possible that the liberal values which increasingly inform this meritocracy have helped fuel the white male political reaction that Wendy Brown sees as the shadow of neoliberalism.

Social Darwinism

Champions of market freedom imagine a terrain upon which economic actors are free to compete with one another without government interference. In this deeply individualistic perspective society doesn't exist – old boys clubs and networks don't exist, oligopolistic cliques and collusions don't exist, the power of giant corporations to suppress competition and distort markets doesn't exist, professional associations don't exist and trade unions are not meant to exist! This is Robinson Crusoe economics, economics as if other human beings (let alone non-human species) don't exist.

So it isn't quite right to say that the neoliberal world is the world of the jungle in which economic actors compete and the fittest survive because of course in the real jungle cooperation occurs both between and within species. Nevertheless it is the myth that is important and the myth is of the heroic entrepreneur, standing on his or her own two feet, fighting off predators, seeking out opportunities and building a business. Read the first part of Ayn Rand's novel *Atlas Shrugged*, the bible of neoliberalism, to see it all there.

We are all paranoid now

The neoliberal subject wanders alone in his hyper-individualised world desperate for company, a monad desperate for the company of other monads. This is a schizoid individualism in which the individual has retreated into himself for outside there be monsters; vigilance is required at all times.[13] So preoccupied is he with avoiding predators that he (conveniently) fails to see his own predatory behaviour, but then it is much easier to see aggression in the other than in oneself. Fortresses appear everywhere upon this terrain – the fortress self, the fortress family and the fortress nation.

The fortress self, which will be considered in detail in the next few chapters, is constantly on the alert, its radar is always scanning the horizon, it is versed in the arts of survivalism – emotional distancing, the management of self in public, putting on a performance.[14] The fortress family appears in a society based upon the survival of the fittest. Here the family not only continues in its role as a refuge, a haven in a heartless world as Christopher Lasch[15] put it, but it becomes the vital incubus for the economic advancement of its members. There is an inward lookingness about this kind of family and, given the real and imagined predators beyond, the family easily takes on a gang-like form.[16] Finally the fortress nation, this is one whose (legally and militarily) protected frontiers are designed to keep out aliens thus securing the nation for 'its own' people. As climate chaos spreads, social collapse deepens, and mass migration increases, many hold that the creation of fortress nations will become a seductive strategy for populist politicians playing on peoples' fears.[17]

The commodification of everything

Oscar Wilde's saying 'people know the price of everything and the value of nothing' usefully captures the way in which the commodification of life has proceeded under neoliberalism. Measurement, calculation, quantification.... whether it is the home care delivered to your old mum or your closely monitored online shopping habits, what counts is what can be counted, and if it can't be counted then it doesn't count.

In fact, this is not quite right; if it can be counted it may still not count enough. In 2017, Deloitte (the same accountancy firm that along with PWC and others introduced performance management to the British public services in the 1980s) produced an economic evaluation of the Great Barrier Reef for the Great Barrier Reef Foundation, supported by the Australia National Bank, in which they estimated the 'asset value' of the reef at $56 billion.[18] Given that at the time the most expensive artwork in the world was Gauguin's 'When Will You Marry' which was sold to a Quatari family for $300 million, this makes the Great Barrier reef worth about 170 Gauguin's. You have to laugh if it wasn't so tragic. Scientists now tell us that even if global average temperature increases were to stop right now, over 90% of existing tropical coral reefs will have died out by 2050.[19] In other words, $56 billion is effectively the write-off value of this priceless piece of nature.

Because we are all drowning by numbers, our grip upon what is real and true becomes increasingly tenuous. We become both cynical *and* credulous.[20] Everything is possible and nothing is true; this is the so-called post-truth world, the polluted tide against which climate science and other truths struggle.

The commodification of the self

As we shall see in the next chapter, in the neoliberal world we are all meant to be economic actors, or at least think as if we were one. We must ensure we are fit (physically and mentally) to compete – lean, flexible and resilient. The self becomes reconstrued as a commodity, how can I add value to myself? There are a whole range of apps and products, from Fitbit to Peleton, to help the self improve its physical fabric. The drive to optimise the value of the self infects the fitness and wellness sectors of the economy. Increasingly mental health services, both publicly and privately provided, have become incorporated to improve the emotional fabric of the self, to enable the self to become more resilient.

Nihilism

The commodification of nature, education, care and the self corresponds to the destruction of their intrinsic value. Probably all of us know this or at least suspect it, that for neoliberalism nothing has value for its own sake any longer. So values gradually lose their value, becoming tradeable commodities like 'greenwash' thus taking the process of disenchantment with the world to its furthest limits. 'Weasel words' and phrases – empowerment, 'making a difference', 'growing the organisation', resilience, sustainability – proliferate and in the process what once had some goodness in it becomes corrupted and twisted.

Wendy Brown asks,

> how might we address the deep, unavowed nihilism and despair of our time? A nihilism that has been growing for a couple of centuries as Nietzsche promised, abetted by neoliberal reason. A nihilism that makes truth and reason into a plaything, that makes values fungible, that vitiates conscience and felt responsibility for the present or the future by the powerful and the powerless alike.[21]

Nihilism is the soil in which the new authoritarians flourish.[22] As we shall see in chapter 11 the Trumps and Johnsons of this world have an utter contempt for truth. Nothing matters beyond their own pleasure. Wendy Brown notes that scandals involving politicians no longer bring outrage but 'a knowing grimace, nihilism's signature'.[23]

---

[1] I use the present tense here even though there is a strong argument to say that neoliberalism has now had its heyday. Much depends upon the extent to which globalization was integral to neoliberalism and therefore whether neoliberalism and nationalism are compatible. Certainly the Biden administration in the USA has reinvigorated the role of the state as a regulator, planner and investor. More typically a free market approach has been combined with reactionary forms of nationalism, as in Bolsanaro's Brazil and Orban's Hungary, leading some to call this neo-illiberalism.

[2] More correctly I should say the 'commodity relationship'. Markets have existed in rudimentary form for millennia and not just across Judeo-Christian civilization. The problem we face concerns a particular form of market economy, the commodity form, which has a built-in logic to accumulate and grow.

[3] Public Sector Economics formed the advanced guard of neoliberalism as it sought to roll back the state. Public sector economists described public officials as 'income maximisers'. Their argument was that public officials had a vested interest in limiting accountability, competition and efficiency thereby giving themselves a cushy life and wasting taxpayer's money. For a clear statement of this approach, one which acted as a catalyst for Thatcherite 'reforms' of the UK welfare state, see Pirie, M. and Butler, E. (1989) *Extending Care*. London: Adam Smith Institute. Note the barefaced deception of the pamphlet's title.

[4] Thatcher, M. (1987) Aids, education and the year 2000! (an interview with Douglas Keay), Women's Own, 31 October 1987. Of course, in many respects Thatcher was an ardent nationalist, something which leads credence to the argument that neoliberalism and nationalism are perfectly compatible.

[5] Wendy Brown provides an extensive exploration of this concept of a neoliberal rationality. Brown, W. (2019) *In the Ruins of Neoliberalism: The Rise of Antidemocratic Politics in the West.* New York: Columbia University Press.

[6] Patmore, C. (2004) Quality in home care for older people: Factors to pay heed to. *Quality in Aging*, 5 (1): 32-40.

[7] I invite the interested reader to do a simple website search under the terms 'mitigation banking' or 'critical natural capital'. You'll be overwhelmed by the exponentially expanding list of entries.

[8] It might be thought that environmentally informed economics would be a good thing but in practice much of it subordinates the environment to economic rationality by adapting economic orthodoxies such as cost-benefit analysis to environmental issues.

[9] Brown, see note 5 above, Chapters 4 and 5.

[10] There has been renewed interest in the role and power of meritocracy in neoliberal society. Whilst Michael Sandel does not link his critique of contemporary American meritocracy explicitly with neoliberalism it is noticeable that his polemic focuses almost entirely on developments since the late 1970s. In contrast Jo Littler makes this link quite explicit in her critique of meritocracy as both an ideology of the 'lonely empowerment' of individual strivers and as a new form of exclusion built upon access to higher education. Sandel, M. (2020) *The Tyranny of Merit: What's Become of the Common Good?* Allen Lane; Littler, J. (2018) *Against Meritocracy: Culture, Power and Myths of Mobility.* London: Routledge. Jo Littler provides a short and clear exposition of her views in 'Meritocracy as Neoliberal Mantra'. See https://archive.discoversociety.org/2018/10/02/meritocracy-as-neoliberal-mantra/ (accessed 30 August 2023).

[11] Kenan Malik argues that the rhetoric of diversity has become integral to today's meritocracy and adds that a diverse elite is still an elite, a socially diverse elite is perfectly compatible with a fundamentally unequal society. Malik, K. (2023) Focusing on diversity means we miss the big picture. Its class that shapes our lives. *Observer*, 29 January, 2023.

[12] Perhaps one can see this meritocratic approach is able to flourish even within the British Conservative Party. The ethic of hard work, personal responsibility, patriarchy and the prioritisation of the family is something which characterises many migrant families. So maybe it should come as no surprise to see the prominent role in British

Conservative politics of figures such as Rishi Sunak, Kwasi Kwarteng, Priti Patel, Nadhim Zahawi, Sajid Javid, Suella Braverman and Alok Sharma all of whom have a second-generation migrant background.

[13] Within psychoanalysis this form of narcissism is often referred to as 'thin skinned'. Bateman, A. (1998) Thick- and thin-skinned organisations and enactment in borderline and narcissistic disorders. *International Journal of Psychoanalysis*, 79: 13-25.

[14] The arts of the survivor will be considered in more depth in Chapter 14.

[15] Lasch, C. (1995) *Haven in a Heartless World*. New York: W.W. Norton.

[16] Some of the first studies of what became known as *amoral familism* were inspired by the fiercely protective and inward looking extended clan families of Southern Italy and Central America. In such societies both state and civil society were rudimentary and the family was responsible both for the protection, welfare and employment of its members, the latter often involving what today we would describe as nepotism. In these traditional societies the family provides a very limited form of social solidarity which may not even extend as far as neighbours. The political scientist and anthropologist Edward Banfield argued that this kind of family system emerged where village communities were unable to generate any sense of common good or purpose. Within the social policy discipline some have suggested that the notion of *amoral familism* closely describes the role of the family under neoliberalism. Rodger, J. (2003) Social solidarity, welfare and post-emotionalism. *Journal of Social Policy*, 32(3): 403-421; Reay, D. (2014) White middle-class families and urban comprehensives: the struggle for social solidarity in an era of amoral familism. *Families, Relationships and Societies*, 3(2): 235-249.

[17] This nationalist 'solution', often referred to as 'the politics of the armed lifeboat', will be considered in detail in Chapter 12.

[18] Deloitte Access Economics (2017) At what price? The economic, social and icon value of the Great Barrier Reef. Available at: https://www2.deloitte.com/content/dam/Deloitte/au/Documents/Economics/deloitte-au-economics-great-barrier-reef-230617.pdf (accessed 5th October 2023)

[19] Zeitvogel, K. (2011) World's coral reefs could be gone by 2052: study. PHYSORG, February 23rd. Available at: https://phys.org/news/2011-02-world-coral-reefs.html (accessed 30th August 2023).

[20] Arendt, H. (2017 [1951]) *The Origins of Totalitarianism*. London: Penguin. p.499 ff.

[21] Wendy Brown (2017) Apocalyptic Populism, https://www.eurozine.com/apocalyptic-populism/ (accessed 30 August 2023).

[22] Brown notes that Irving Kristol, the darling of North American neoliberal thinktanks, offered two cheers for capitalism but saw consumer societies as empty of moral meaning 'if not forthrightly nihilistic'. According to Kristol, an explicitly conservative moral political programme was required to counter this effect. This led to a 'transactional' engagement between (largely atheist) libertarians and conservative evangelicals - a particularly striking feature of the collaboration between the amoral Trump and the religious right in the period between 2016 and 2020. Brown, see note 5 above, Chapter 3.

[23] Ibid, p.162.

CHAPTER FIVE

THEY WANT ALL OF YOU

Self-optimisation

After the Second World War in Western economies manual workers were steadily replaced by managerial, professional and service workers as digital, cultural and service sectors superseded manufacture as the engine of growth in post-industrial societies. Post-industrial society is powered by cognitive, communicative and interactive skills. As several writers[1] have noted, these new forms of work pose new challenges relating to the management of the more educated and feminised labour force. New modes of control have emerged which are concerned to harness the educated employee's desire not just to do well but to achieve excellence,[2] that is to be the Ideal.

In the modern organisation the self, like the environment, has become a resource to be managed so that value can be extracted from it. By the 1990s Human Resource Management (which does what it says on the tin) had become concerned to optimise the performance of the self through a range of 'performance management' techniques. The resulting systems of targets, indicators, comparators, benchmarks, 360 degree feedback, etc. have come to dominate both private corporations and government.[3]

Today we are all meant to be economic actors, or at least think as if we were one. We must ensure we are fit (physically and mentally) to compete – lean, flexible and resilient.[4] The self becomes reconstrued as a commodity, how can I add value to myself? There are a whole range of apps and products, from Fitbit to Peleton, to help the self improve its physical fabric and hone its achievement orientation. Deborah Lupton saw Fitbit as a harbinger of what she called 'self tracking communities' around fitness centres and websites, such as Quantified Self, where subscribers shared downloaded Fitbit data and 'benchmarked' their performances against peers.[5] Self-trackers engaged in a constant process of monitoring their vital signs (pulse, blood pressure, blood oxygen saturation levels, etc) as they ran, pushed weights, chilled, worked and slept. Lupton cites one self-tracker saying 'my goal is to basically be – an optimal human being in every aspect of my life'.

The modern organization wants all of us; it wants our personality, our flexibility and our creativity. To stay ahead of our competitors, we need to engage in constant self development. It is not just financial markets that operate in a constant, 24-hour, real-time flow but, as William Davies noted, with their smart phones, iPads, etc. an increasing number of today's professionals 'now dwell in the same state of constant anxious fluidity as a financial trader or entrepreneur'.[6]

Performance relates to the world of excellence, of 'doing well', mastery, pride, striving. Some refer to this as psychological self-optimization, 'this is a vision of psychological life as enterprise, one centred on the individual pursuit of wellbeing as one of calculating self-interest'.[7] Here one constantly measures oneself against a set of external criteria – am I meeting my targets? Performativity is about doing not being, it conjures that imagined world in which (according to the tired jingles of that army of HR managers) the 'empowered' self goes about 'making a difference'. It is the world of the active voice, of imaginary control over one's environment and over one's own self, of shaping one's destiny, taking command of one's 'journey'.

Performance anxiety

Caught in the grip of the Ideal, the modern self hardly ever feels itself to be enough, it never measures up, it always falls short before the imagined gaze of the Other. Brief moments when the demand of the Ideal is satisfied punctuate the background hum of performance anxiety, they mostly bring relief rather than euphoria. Fitbit's self-trackers simply give shape and form to a self which is engaged in a treadmill-like way not just with work but with all areas of life including leisure and consumption, spheres where we are told we are free. Modern workers are in a struggle for psychological survival, they are members of what we might think of as a psychological precariat.

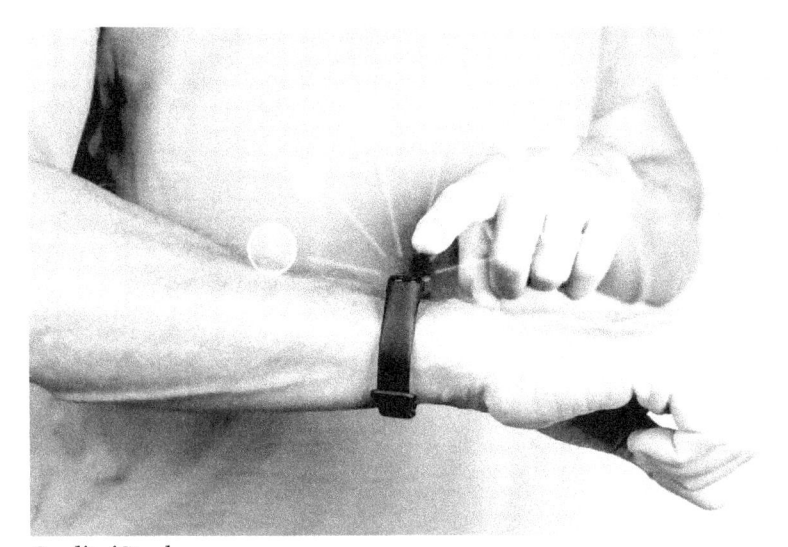

Credit: iStock

Because the fragile self is so preoccupied with everyday psychological survival a realistic awareness of external dangers is absent other than in fleeting and catastrophising ways which make effective action impossible. Survivalism is best illustrated not by the activities of religious cranks holing up in deepest Montana as they wait for the rapture nor, more benignly, by those who drop out to build ecotopias in the few wild places left in the world. Rather survivalism[8] best describes the internal climate and everyday life of the majority of neoliberal citizens, and not just in the West but increasingly in China and other 'modernising' countries.

Survivalism is therefore first and foremost a state of mind,[9] a way of being in the world. In my experience, which may of course be specific to living and working as a therapist in Europe, this state of mind often draws upon the cultural legacy of the Holocaust. To give an example, a patient of mine, a middle manager in a public organisation, was suffering from ill health brought on by her utter exhaustion. When I expressed curiosity about why she didn't ask for sick leave her reply was to the point: "because I fear I'll be dispensed with". She often had dreams about being interned in a concentration camp and her use of the phrase 'dispensed with' clearly resonated with this.

The paranoid dimension to the survivalist state of mind is clear. This woman's employer had a reputation for incompetence but not for ruthlessness. But like all large modern organisations, its performance management systems sank deep into the everyday lives of its employees where everything was subject to quantification and measurement. And of course this organisation was situated in a wider cultural context where genuine vulnerability[10] was regarded as a possible indication of personal inadequacy.

People have to struggle through the persecutory experiences of performance anxiety before they can get to the experience of loss (particularly the sense of a life passing) and guilt. This in turn is often so painful that it leads to a flight back into survivalism. I wonder whether the fragility of the self in neoliberal society is such that an encounter with the reality of climate change and the loss, grief and guilt that this entails, would actually be a significant cultural achievement, equivalent to a radical shift from a survivalist culture to what Sally Weintrobe calls a 'culture of care'.[11]

---

[1] Hochschild, A. (1983) *The Managed Heart: Commercialisation of Human Feeling.* Berkeley: University of California Press; Boltanski, L. and Chiapell, E. (2007) *The New Spirit of Capitalism.* London: Verso.

[2] The classic management text of this ilk, *In Search of Excellence*, appeared in the early 1980s just as neoliberal governance was getting going. Peters, T. and Waterman, R. (2015 [1982]) *In Search of Excellence*. London: Profile Books. It was the most widely owned monograph in the USA between 1989 and 2006.

[3] Reflecting on the educated workforce, the building blocks of today's meritocracy, Stephen Ball noted the 'sense of being constantly judged in different ways, by different means, according to different criteria, through different agents and agencies…. We become ontologically insecure: unsure whether we are doing enough, doing the right thing, doing as much as others, or as well as others, constantly looking to improve, to be better, to be excellent'. Ball, S. (2003) The teacher's soul and the terrors of performativity. *Journal of Education Policy*, 18(2): p. 217.

[4] Privately provided counselling and psychotherapy has become a key part of this wellbeing sector and therefore inevitably infected by this neoliberal narrative of the self. As an example see the critique of the way in which the idea of 'resilience' has been degraded and abused see Evans, B. and Reid, J. (2014*) Resilient Life: The Art of Living Dangerously*. Cambridge: Polity.

[5] Lupton, D. (2016) *The Quantified Self: A Sociology of Self-Tracking*. Cambridge: Polity.

[6] Davies, W. (2015) The chronic social: Relations of control within and without neoliberalism. *New Formations*, 84/5: 40-57.

[7] Binkley, S. (2011) Psychological life as enterprise: Social practice and the government of neo-liberal interiority, *History of the Human Sciences*, 24(3): p. 94

[8] Christopher Lasch was one of the first to chart the emergence of the survivalist mentality in the modern corporation and the wider culture. Lasch, C. (1984) *The Minimal Self: Psychic Survival in Troubled Times*. New York: W.W.Norton & Co. See especially Chapter 2.

[9] A 'state of mind' refers to a mode of organising and processing emotional experience. I provided a detailed account of this concept, largely derived from the work of the psychoanalyst Thomas Ogden, in an earlier book. Hoggett, P. (1992) *Partisans in an Uncertain World: The Psychoanalysis of Engagement.* London: Free Association Books. pp.77-79.

[10] In using the term 'genuine' here I am referring to something which is typically involuntary in contrast to performative vulnerability, a highly prized manifestation of the commodified self, practiced by many celebrities and politicians, including Tony Blair. This isn't a superficial performance but requires what Arlie Hochschild (see note 89) calls 'deep acting'; Hoggett, P. (2005) Iraq: Blair's Mission Impossible, *British Journal of Politics and International Relations*, 7: 418-428.

[11] Weintrobe, S. (2021) *Psychological Roots of the Climate Crisis: Neoliberal Exceptionalism and the Culture of Uncare.* New York & London: Bloomsbury Academic.

EXCURSION

BY THE BANKS OF THE M49

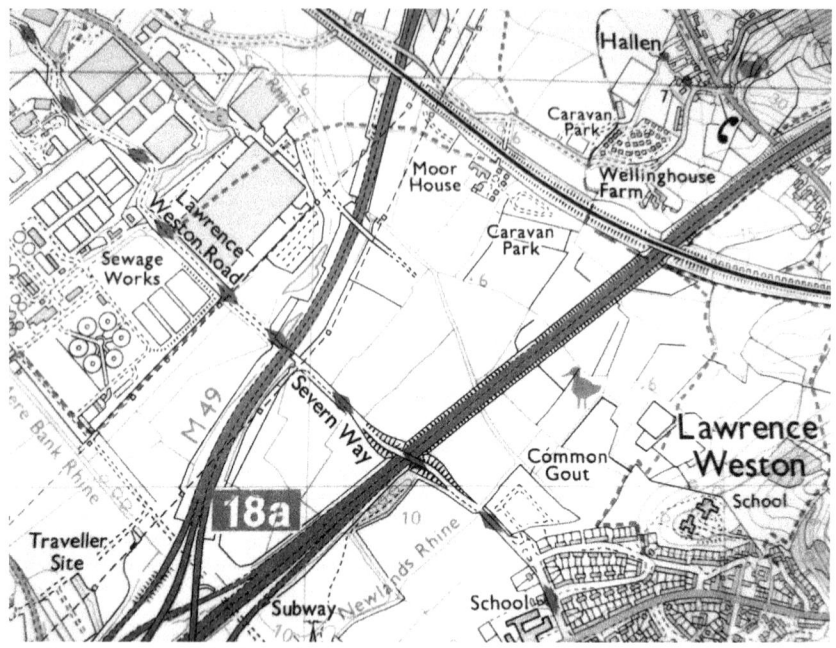

OS©Crown copyright (2010) FL-CS0000863328

Over the ridge where I live you come down onto the floodplain of the Severn. The first place you encounter is Lawrence Weston, a post-war public housing estate wedged between the ridge and the M5 motorway. Cross the M5 and you're in V-shaped area of land between the M5 and the (fairly new) M49 which converge at an intersection at Avonmouth. Move further west and cross the M49 and you're in an industrial area which occupies the land running north from Avonmouth to the New Severn Bridge which crosses into Wales. This area comprises the Avonmouth Docks, several major distribution centres, industrial estates, a Biomass power station, and several large wind turbines. The largest land-based wind turbine in the UK has just been built here as part of a community energy project run by Ambition Lawrence Weston, a brilliant local resident's group. I think the planners have done a pretty good job and in the whole area there are many artificially created wildlife corridors, ditches and retention ponds which I like to explore in the summer. This is Richard Mabey's 'unofficial countryside'.[1]

I find the V-shaped land in between the two motorways particularly fascinating. As the V between the two motorways opens out there's a solar farm, then several well maintained but unused fields, another solar farm and a caravan site beyond which is the village of Hallen. It's probably a distance of two miles from the motorway intersection to Hallen. The Henbury Loop railway line crosses the northern edge of the area and takes goods traffic to Avonmouth. Two country lanes also cross the V. They are now inaccessible to vehicles but regularly used by cyclists commuting back and forth to the distribution centres.

Ok, so there's a lot of traffic noise, but for me it only adds to the wildness of the place. There's a walk I'm particularly fond of. It starts in the village of Hallen. You take Moorhouse Lane, one of the two I've mentioned, from Hallen and head south and then west. First you pass the terrific Hi Ways residential park whose 40-odd mobile homes with gardens are all immaculately maintained. Next comes Hallen's new sports centre and village hall with cricket pitch, all weather surface football pitch and Hallen FC stadium (an amazing complex given the tiny size of the village). Just before the tarmacked road bends left under the Loop Line and into the caravan park, what's left of the old country lane continues on towards the Severn. Now abandoned to all but walkers, cyclists and horse riders the lane is bounded on each side by rhines (the name given to the old drainage ditches). The lane becomes increasingly overgrown until suddenly it opens out into a splendidly wide road bridge crossing the M49. In the late summer there's a profusion of wild plum, sloe and blackberry bushes along the bridge's two inclines. I've put a camera trap on the bridge and have found that it is used by foxes crossing from one side of the motorway to the other. If you look south-west from the bridge towards Avonmouth you'll notice two more bridges follow in swift succession. The first carries the Henbury Loop railway line and the second is the bridge to nowhere which I'll mention later.

Having crossed over the motorway the lane bears left, goes under the railway line and funnels into the complex of distribution centres. After 100 yards there's a path on your left that takes you over a little wooden bridge that crosses one of the rhines and leads into a marshy area where grebes and other wildfowl can often be heard. Although industrial development presses in, the path has been preserved and loops back round towards the motorway. A massive Amazon distribution centre looms in front of you but if you persevere the path twists back and forth until it emerges upon the western incline of the bridge to nowhere. A quite amazing bridge with two wide road lanes and ample footways it emerges out of the marshy scrub next to the Amazon, sweeps over the M49 and deposits you in a large grassy *field. This is one of the several well maintained fields between the two solar farms that I mentioned earlier. This bridge must have cost millions to build but today its main function seems to be to offer a challenge to local Buddleia – one magnificent rounded bush on the bridge has a circumference approximating 16 yards.*

I love Buddleia, to me it represents something about nature in this whole area - it's capacity to endure, improvise and fight back. Now I'm not a flora man but I have treated myself to Picture This, the plant identification app, and I could recite a list of umpteen different types of thistle and wild grasses to be found along my way but I'm afraid I only tend to notice the 'pretty' ones which I can tell you include wild mallow, shortpod mustard, common fleabane and red bistort.

When I'm full of troubling feelings about nature's vulnerability in the face of climate change this is where I like to come, it makes me feel hopeful. I love to stand on the bridge to nowhere, to stand in a place where perhaps no other human has stood in the last few weeks, and watch the traffic busily on its way from South West England to South Wales.

Buddleia on the bridge to nowhere

---

[1] Mabey, R. (1973) *The Unofficial Countryside.* London: Collins.

CHAPTER SIX

TIME IS THE ENEMY

We are human creatures at war with our creatureliness. But what does it mean to say this? An increasingly common view within circles linked to deep adaptation[1] and climate psychology[2] is to say that we are at war with the idea of our mortality and death, that is, with our creaturely nature. If only we could come to accept our own mortality, frailty and transience then we could overcome our separation from nature and learn to respect its own fragility and limits. Indeed the encounter with death and our flight from it has been a recurring theme in twentieth century philosophy,[3] social theory[4] and psychoanalysis.[5] But in my clinical practice I have found that the fear of death rarely has a visceral presence. Sometimes it seems to be projected outwards into a fear for the mortality of loved ones and occasionally it is clearly manifest in hypochondria and the acute anxiety which often accompanies this. But by and large people seem to be 'philosophical' about death; in a way similar to their reaction to climate change, they know about it and can think about it but somehow or another their thoughts are split off from any feeling.

The trouble with time

In contrast, I have found that many of the people I have met in my psychotherapy practice have been deeply troubled by time. For several of them time seemed to be inextricably linked to loss, as in the expression 'time marches on'. Julie, one of my younger patients, was trying to make it as a DJ and music producer. She was terribly anxious about time and life passing her by. Although only thirty, she spoke as if she was already old and 'going downhill'. If for any reason, such as illness, she could not focus whole-heartedly on her music she would panic and this fed a fury with the world which she felt was conspiring against her. She once described cycling to her studio in the rain and muttering 'fuck you' at the wind, at the rain and almost every passing car, as if everything and everyone was out to get in her way, make life difficult for her, hold her back.

Sometimes the link with loss assumes a different form. When the holidays loomed Wes, a family man, would enter a prolonged state of agitation. Where should they go, should they go somewhere they'd already been to and liked or somewhere new, to a small hotel or self-catering, should they opt for chilling by the beach or something more active? If he found himself moving in his mind towards the small hotel, he would suddenly find himself besieged by memories of the beautiful raucous chaos of self-catering. He would drive his family crazy with his procrastination, an agonizing process sometimes lasting months in which different choices were entertained, discounted, revisited, discounted once more and so on. Time and choice go together and choices persecuted Wes because to choose A often meant not to choose B and this then confronted him with loss, the loss of the path not chosen, and thus the loss of the illusion that he could have everything. By putting decisions off he pursued his magical belief that somehow things would work themselves out and he could both have his cake and eat it. But of course this never happened and he would be forced to make decisions at the very last minute.

Time haunts people such as Julie and Wes, it stalks them, tantalizes them and sometimes terrifies them. Another patient of mine, Stuart, took his grandfather, a successful military man, as his ideal. One of his grandfather's sayings was 'time is the enemy'. Like Julie, Stuart was only in his thirties but at times he spoke as if his best times were already behind him.

The fear of time can elicit some novel responses: Philip feels a huge sense of relief when he gets home from a session with me and knows that there is now a whole week without any further demands on his time (he has been unemployed for over four years). If he can come ten minutes late and blame it on the buses, he has a sense of triumph that he has squeezed a little more time for himself and cheated me of 'my time'. He hoards time like a scarce currency, delaying sleep for as long as he can; he hates it when, having got engrossed in a movie he looks at his watch and says 'gosh, is that the time!'. The point of having time for Philip is not to do something with it, it is not a means to an end, it is an end in itself. The ideal is to have endless time, all the time in the world, and to have to do nothing and then to do nothing. Philip once said, 'I would probably say "I just want to chill" whereas the reality is I want to stall time.'

Modern civilization has ensured that mortality and death are largely absent from our lives. These things largely occur in hospitals, nursing and care homes and hospices, safely 'out of sight and out of mind' for the rest of society. In contrast time bombards us with its presence and persecutes us with its constant passing. As the psychoanalyst Henry Smith put it, 'if mortality is our ultimate enemy, then time is its henchman'.[6] Time is loss, time is money.

David Bell suggests that the capacity to exist in time is a developmental achievement but for some people time 'is felt as a fixed, imminent catastrophe to be evaded by the creation of a timeless world' (p. 784).[7] Philip's attempt to 'stall time' vividly illustrates this attempt to create a timeless world. On the other hand, Julie's 'panic' suggests someone who is still stalked by the imminent catastrophe of time. Wes's chronic indecision,

PARADISE LOST? The Climate Crisis and the Human Condition   53

which was also a characteristic of one of Bell's patients, reveals the connection between the flight from time and the illusion of an ideal world where it is possible to have everything.

You might say that in Paradise time stands still if it weren't for the fact that in Paradise there is actually no such thing as time.[8] A timeless world is an ideal world, a world free from loss, lack and privation, it is one that contemporary society tells us we are entitled to. Sally Weintrobe connects this to exceptionalism. Exceptions imagine that the world they omnipotently construct and control abolishes time, for the true exception: '[T]he clocks are stopped'.[9] In a narcissistic society such as ours, the ideal world is a world without time.

No speed limit

It seems as if we are either fighting time or in flight from time. Culturally our flight appears to be manifest in a search for distraction, a kind of collective attention deficit disorder. Surely we have all experienced this, you're sat on the train or the bus and everyone has their head buried in their mobile device as the world beyond passes them by. I'm struck by the huge paradox that whilst our contemporary society is preoccupied with the 'now' of things – what is the latest happening right now, what is the latest trend, innovation, podcast, livestream, tweet – many of its citizens nevertheless appear to have huge difficulty living in the present. Increasingly we seem to live inside our heads. Being present in the actual physical world or just 'being with' an other becomes an extraordinary challenge.

To the extent that we are oppressed by our internal nature we are oppressed by time. For time reminds us of life passing, of things not done, or of things done but now gone, lost. For some people I know this oppression is almost unbearable, as if all that they are in touch with is life's absence. Some, like Julie, are completely focused on the future; others, like Stuart, are consumed by ruminations about the past: 'life goes on elsewhere, but not here, not in my life.' In my consulting room they hear my clock ticking and it persecutes them.

There is a societal echo of this. In the modern world time is intolerable and must be compressed, accelerated, annihilated. Nowhere is this more true than in the organisation of globalised business.[10] Since the Second World War humanity's productive forces have been unleashed to such an extent that we do indeed stand just as Prometheus once did.[11] And far from slowing down everything is speeding up.

The idea of 'accelerationism' has become popular recently,[12] it refers to several connected features of contemporary economic and cultural life. The confluence of unrestrained capital and the development of information technologies, AI and robotics has created a runaway world in which (to cite the title of Steven Shaviro's recent book)[13] there seems to be no speed limit any longer in life, neither in economic life, nor in cultural life, nor in private life, nor in carbon emissions. This is Joseph Schumpeter's 'creative destruction'[14] gone mad. Faster, faster…. Time – mortality's representative and henchman – is the enemy. It must be overcome, it must be compressed, frozen, reduced and eradicated.

I sometimes fear that the world is currently run by those who are afraid of living. Their fear drives them on. It's an itch, an ache, a restlessness. Always in motion, always overcoming, always the next thing, always the new. They long to be free of longing, but they don't know how. And so they rise to the top, of business, government, art and literature and they shape our world. Our world, the world of the contented, of the ordinarily happy, of the just going on being.

1 Bendell, J. (2021) Co-liberation from the ideology of E-S-C-A-P-E. In Bendell, J. and Read, R. (eds) *Deep Adaptation: Navigating the Realities of Climate Chaos*. Cambridge: Polity Press.
2 Gillespie, S. (2020) *Climate Crisis and Consciousness: Re-imagining Our World and Ourselves*. Oxon & New York: Routledge. Chapter 3.
3 One of the few attempts I have come across to examine the connections between existential philosophy and climate change has been undertaken by Ruth Irwin. Irwin, R. (2008) *Heidegger, Politics and Climate Change*. London: Continuum.
4 The classic text here is undoubtedly Ernest Becker's *The Denial of Death*. Becker, E. (1973) *The Denial of Death*. New York: The Free Press.
5 Becker argues that Freud found the prospect of his own mortality very troubling and death as a concrete reality rather than abstract force (Thanatos) is largely absent from both his theory and recommendations on practice, unlike Otto Rank for whom it was pretty central. See Becker, *Denial of Death*, note 4 above, pp.97-105.
6 Smith, H. (2016) Time and Mortality, *Psychoanalytic Inquiry*, 36: 420-434.
7 Bell, D. (2006) Existence in time: Development or catastrophe. *The Psychoanalytic Quarterly*, 75: 783-805.
8 Steiner, J. (2018) Time and the Garden of Eden Illusion. *International Journal of Psychoanalysis*, 99 (6): 1274-1287.
9 Weintrobe, S. (2021) *Psychological Roots of the Climate Crisis: Neoliberal Exceptionalism and the Culture of Uncare*. New York & London: Bloomsbury Academic. p.48.
10 See for example Hassan, R. and Purser, R. (eds.) (2007) *24/7: Time and Temporality in the Network Society*. Stanford, CA: Stanford University Press.
11 In Greek mythology Prometheus steals fire from the Gods, to have Promethean power is therefore to be God-like.
12 Noys, B. (2014) *Malign Velocities: Accelerationism & Capitalism*. Winchester: Zero.
13 Shaviro, S. (2015) *No Speed Limit: Three Essays on Accelerationism*. Minneapolis: University of Minnesota Press.
14 The idea of 'creative destruction' is commonly linked to the postwar Austrian economist Joseph Schumpeter. Schumpeter drew upon Marx when he drew attention to the way in which capitalism constantly disrupts economic life via waves of growth associated with the business cycle. The restless creation of the new is accompanied by the inexorable destruction of the old – old ways of doing things, old forms of life. Schumpeter, J. (1994 [1942]) *Capitalism, Socialism, Democracy*. London: Routledge.

CHAPTER SEVEN

SHAME

The feeling of failure

As neoliberalism weakened the solidarity that bound workers, citizens and communities together in the post war period, the experience of failure became increasingly individualised. I began to work as a psychotherapist during the first years of the twenty-first century and my societal antennae began to pick up traces of this in the lives of the patients that I started to see. My psychoanalytic training had prepared me to expect people burdened by guilt but what I found was that those coming through the door of my consulting room were typically overwhelmed by shame. They felt that they were simply not good enough, and to be not enough was to be a failure, a loser, a nothing.

Shame is the emotion of failure.[1] Unlike guilt, which relates to what we do, shame relates to who we are and specifically to the feeling that we are inadequate, deficient and 'not enough'. Melvin Lansky, the psychoanalyst who probably more than any other has helped reorientate psychoanalysis towards shame connects it to the 'awareness of failure to meet standards and ideals, from exposure as inadequate or deficient'.[2] Remember what I said in Chapter 3 about The Law of All or Nothing, under the tyranny exercised by the Ideal if I cannot be the Ideal then I am nothing.

Freud thought guilt was endemic to society and to the psychic lives of his patients. Nowadays an increasing number of therapists[3] believe that guilt has been superseded by the shame which haunts the fragile or thin-skinned narcissism of the neoliberal citizen. According to Freud conscience and guilt were the mechanisms through which the moral order of society was sustained.[4] If life is not to be nasty, brutish and short, then civilization demanded that we must accept restrictions upon impulse and desire, particularly on aggression. Guilt was the manifestation of this moral inhibition within the psyche.

However, by the second half of the twentieth century, as a substantial professional-middle class emerged, modern Western society was changing from a scarcity-based society to one of relative affluence in which economic growth required constantly expanding horizons of consumption. The evolution of psychoanalysis during the twentieth century reflected the shift from the restrictive society of Freud's time to the permissive society of the postwar years.[5] There was a shift from a society based upon relative scarcity, where desire was to be disciplined by the inner-directed controls of conscience and guilt, towards a permissive society based upon relative wealth and abundance, where desire was constantly shaped and cultivated via the mass and social media and the legions of commercial taste makers. What is my desire? My desire is the desire of the Other.[6] David Reisman termed this 'other-directedness',[7] and it is shame not guilt that powers it.[8]

Fear of shame fuels the performative regimes which now envelop the lives of today's workers and consumers, particularly of the new meritocracy. Whilst winners are idealised, no-one wants to be 'a loser'. If we cannot be the Ideal then we are a failure. As we have just seen, workers in new sectors such as the media, digital technologies, science and human services tend to identify with their work and want to do well at it.

But in a system where success is idealised this identification becomes an Achilles' heel as it is harnessed and exploited by employers who always want more from their workforce. A gnawing, semi-permanent feeling of failure and inadequacy is the consequence of such performative regimes where the ideal of success remains forever elusive. Under neoliberalism both the new workforce and the traditional, manual workforce have been subject to the same processes of downsizing, casualization and flexibilization. The consequence is a survivalist culture in which performance anxieties reach right down through the body and into the soul.[9]

The 'abject' self

Reflecting on the entrepreneurial self of neoliberalism, Lynne Layton noted the accompanying denigration (and shamefulness) attached to vulnerability. She saw the weakening/fragmenting of social ties increasing the narcissistic wounds arising from the sense of failure and isolation.[10] This 'abject self' manifests a fragile narcissism which is far from being grandiose. There is little euphoria when objectives are met, just temporary relief from the otherwise ongoing feelings of persecution. The fragility of this 'thin skinned' narcissism is manifest in the constant vigilance in relation to almost invisible slights and setbacks that precipitate narcissistic injuries, something I will explore in more depth in the next chapter.

Credit: Pixabay

It is the intensely performative character of neoliberal society that provides the cultural underpinning for this fragile narcissism. This performativity has two different facets. The first, discussed already, resembles 'the management of the self by objectives' and concerns the constant monitoring of self's achievements against standards and targets. The second refers to the dramaturgical meaning of performance, captured by Goffman's phrase 'the presentation of self in everyday life'.[11] Arlie Hochschild develops Goffman's analysis of everyday performance by making an important distinction between deep and surface acting.[12] In the latter case a role, such as being cool, has become so well rehearsed that the actor is no longer aware that they are acting the part. They have taken the role into themselves and thus the distinction between true and false self has broken down. Don DeLillo captures this in the following passage from *Underworld*:

> I noticed how people played at being executives while actually holding executive positions. Did I do this myself? You maintain a shifting distance between yourself and your job. There's a self-conscious space, a sense of formal play that is a sort of arrested panic, and maybe you show it in a forced gesture or a ritual clearing of the throat. Something out of

childhood whistles through this space, a sense of games and half-made selves, but it's not that you are pretending to be someone else. You're pretending to be exactly who you are. That's the curious thing.[13]

Here we encounter that exaggerated form of other-directedness in which the self only feels it exists in the eyes of the other. Two of my younger patients referred to this in terms of entering a state of 'torpor' when close friends on whom they depended for their sense of self went away. Another, a very successful senior executive, was terrorised by the thought that if someone got close to him they would discover there was nothing there beyond his public persona. He described himself as a 'man without substance'. At the cultural level Richard Sennett[14] argued that this preoccupation with how one is seen in the context of the new flexible and performative organisational cultures corresponds to a culture in which a more superficial sense of 'personality' has replaced the deeper idea of 'character'.

The abject self is the psychical counterpart to a society increasingly dominated by a ruthlessly competitive, individualised and precarious labour market, in which social safety nets have been pared down or stripped away, and where social ties and bonds (e.g. family, community, public service) not based upon market principles are progressively corroded and undermined. This self is engaged in a struggle for survival which is as much real as it is imaginary.

But the abject or minimal self seems to be only one part of the story; there is also a sense of entitlement. The splitting of the lifeworlds of work and consumption during the neoliberal period parallels the internal split between the abject and entitled self. The world of consumption acts to legitimise the experience of unfreedom in the world of work.[15] The entitled part of the self wants it and wants it now. The entitled self is intolerant of loss and doesn't see why s/he should have to give up anything. This is captured in some of Robert Tollemache's interviews about flying. As one respondent remarked,

> that is a difficult one, we love travelling, jump on a plane, you know, off we go, we do it a lot………Yeah, [wife's name] and myself, you know, we like, we like to travel; we went to India at Christmas, because we can; (he laughs) we did, let alone what that means, what that nine hours of plane in the sky there and nine hours back meant for the environment we didn't consider. Even though we are both intelligent enough to know that, lots of other people doing that is probably not sustainable.[16]

The phrase 'because we can' captures this unconscious sense of entitlement perfectly, it cultivates the myth that choice does not involve loss – one can choose A *and* choose B.[17] The entitled self almost seems offended by the presence on Greek holiday beaches of the displaced migrant who bears the cost of his affluence.

Shame-based cultures are not new, indeed they are as old as antiquity. In traditional societies shame arose from the experience of dishonour and disrespect in relation to one's group. But in modern societies the cause of shame shifts from dishonour and disrespect in relation to one's group towards failure and weakness in relation to one's self. Moreover the shaming no longer emanates from an identifiable community but from an anonymous 'public', the Other.

It is striking how pervasive shame has become in everyday life. John Ronson, whose book *So You've Been Publicly Shamed* looks at on-line shaming, argues that we are living through a great renaissance of public shaming.[18] There is a kind of mob mentality here and a terrific self-righteousness. At a more intimate level there seems to be an epidemic of online shaming in schools using Facebook and Instagram - from 'slut shaming' to the various forms of 'cyber bullying' now reported right across Europe and the US.

Unsurprisingly there is now evidence of a direct correlation between body shame and the use of social media such as Facebook.[19] Body shame is now rife among younger people in Europe and the USA; indeed there is now a massive research industry devoted to this subject. The internalisation of the persecuting gaze of the Other finds expression in concepts such as 'body surveillance' and 'self objectification'.[20]

Shame and shaming

Shame is a profoundly painful feeling and nearly all commentators agree that shame lends itself easily to projection and reversal.[21] One avoids shame in oneself or one's group by locating it in the other – 'I might not be educated or have money but I am white and proud of that' or 'I might be black but at least I'm a real man not like some of those effeminate gay men'. By regarding the other with contempt some of the rents and fractures in one's own fragile sense of self can be temporally made good. But this comes at a cost and the cost is the fracturing of the solidarities that might otherwise unite those without wealth and/or power. My anxiety is that the new liberal meritocracy, haunted by performance anxiety, deals with the shame that threatens it by projecting it onto others.

In many western democracies, classism is now a powerful cultural phenomenon. White working class people are often derided by the meritocracy for what is seen as their parochialism, their poor taste in food, music,

PARADISE LOST? The Climate Crisis and the Human Condition 61

TV etc, for the way they speak and dress, for how they vote, for how little they know about the world, and this scorn is conveyed just as powerfully in liberal media as in conservative. This is class shaming.[22] There is a grave danger that the liberal meritocracy, which provides the backbone of environmental movements in countries like Britain and the USA, contributes to the shame and resentment which currently fuels reactionary forms of populism.

Some have argued that eco-anxiety is a primarily white and middle class anxiety?[23] The prospect of an austere and precarious future may strike panic into the middle classes of Britain and the USA. But such precariousness is now, and has been in the past, the lived experience of huge sections of the population living within our own countries, let alone the rest of the world. This is an experience many environmentally minded activists seem divorced from and uncurious about. This blindness is both racist and classist, and whilst the climate change movement may be awakening to the former, I believe that it is still largely deaf and blind to the latter.

The weaponization of shame

Writing just before Trump's victory in the presidential election Adam Haslett[24] argued that Donald Trump had weaponised shame. Haslett argued that shame was now a pervasive condition in American society, one endured by women, black people and other minorities as well as the forgotten and dispossessed white working class. Trump gave people the means to avoid the shame of their condition, what Haslett calls 'the shame of being lesser-than', by enjoying, live or online, his shaming of others. I would add that one of the striking characteristics of Trump is his shamelessness, indeed one cannot think of anything – groping women, insulting the disabled, not paying taxes – that could cause this man shame and perhaps here lies the origin of the aura of invincibility that surrounds him.

The response of the liberal left to white working class populism such as Trumpism in the USA and Brexiters in the UK has often been contemptuous. How could they be so stupid as to vote, like turkeys, for Christmas? But by questioning the intelligence of the white working class, left behind by the de-industrialisation of advanced western economies, liberal progressives contribute to the ongoing shaming of this social group. The white working class are treated with contempt and become easy prey both for authoritarian populists and for the continuation of the neoliberal social policies which made society's poorest pay the price for the financial crash of 2008.[25] Such policies would, if allowed to continue, also force the least well off to shoulder most of the cost of the implementation of policies to mitigate climate change. Rather than project our shame upon those we imagine are somehow below us, how much more effective it would be to understand, as Haslett says, that 'shame is what we have in common'.

---

[1] The classic early text on shame was by Helen Bloch Lewis. More recently Melvin Lansky and Leon Wurmser have made important contributions. Lewis, H.B. (1971) *Shame and Guilt in Neurosis.* New York: International Universities Press; Lansky, M. (1999) Shame and the idea of a central affect. *Psychoanalytic Inquiry*, 19: 347-361. Wurmser, L. (1981) *The Mask of Shame.* London: John Hopkins University Press.
[2] Lansky, ibid. p.351
[3] The Kleinian tradition within psychoanalysis has been slow to catch up with the growing interest in shame although John Steiner's work is an important recent exception. Steiner, J. (2011) *Seeing and Being Seen: Emerging from a Psychic Retreat.* London: Routledge.
[4] Freud, S. (1929) Civilization and its Discontents. In J. Strachey (Ed.) The Standard Edition of the Complete Psychological Works of Sigmund Freud, Vol 21, p.67.. London: Hogarth Press.
[5] Giddens, A. (1991) *Modernity and Self-Identity: Self and Society in the Late Modern Age*. Cambridge: Polity. pp. 153-5
[6] In using the capital 'O' here I'm implying that the other in question is not so much a real, concretely existing other like one's best mate or a neighbour, but a fantasised figure invested with enormous significance and power. One of my patients was struggling to make his name in the fashion industry and was literally terrified of those he called 'the scaries'. The scaries embodied both 'cool' and high achievement, they represented all that he felt he should be (but felt he wasn't). Occasionally he would meet a real scary and would discover, to his surprise, that they were just ordinary people with the same hopes and worries as himself but in no way did this encounter with reality weaken the power of his fantasy.
[7] Riesman, D., Glazer, N. and Denney, R. (1950) *The Lonely Crowd: A Study of the Changing American Character*. New Haven: Yale University Press.
[8] Lasch, C. (1979) *The Culture of Narcissism: American Life in an Age of Diminishing Expectations*. Ney York: W.W. Norton. Lasch (p.63) acknowledges his debt to David Riesman, the culture of narcissism provides the social context in which Riesman's other-directed character flourishes. Lasch's analysis of the shifting nature of internalised controls is very similar to the one I have presented here. As he says, '(T)he changing conditions of family life lead not so much to a 'decline of the super-ego' as to an alteration of its contents…the development of a harsh and punitive super-ego… holds up to the ego an exalted standard of fame and success and condemns it with savage ferocity when it falls short of that standard' (p.178).
[9] The emergence of a new section of the workforce who are self-employed but largely not out of choice - the delivery drivers, couriers, agency workers, the young workers in the gig economy – has led some to refer to them as 'the precariat'. Standing, G. (2011) *The Precariat: The New Dangerous Class*. London: Bloomsbury Academic. What I'm suggesting is that alongside the precariat of the gig economy a new psychological

precariat has emerged within the educated meritocracy, the result of the performative regimes which have gripped organisations under neoliberalism.

[10] Layton, L. (2014) Some psychic effects of neoliberalism: Narcissism, disavowal, perversion. *Psychoanalysis, Culture & Society*, 19(2): p. 166.

[11] Goffman, E. (1959) *The Presentation of Self in Everyday Life*. New York: Anchor Books.

[12] Hochschild, A. (1983) *The Managed Heart: Commercialisation of Human Feeling*. Berkeley: University of California Press.

[13] DeLillo, D. (1997) *Underworld*. New York: Simon & Schuster.

[14] Sennett, R. (1998) *The Corrosion of Character: The Personal Consequences of Work in the New Capitalism*. New York: W.W. Norton & Co.

[15] In conversation with Mark Fisher, Jeremy Gilbert notes that the splitting apart of the two worlds of work and consumption creates 'a mode of subjectivity which ultimately accords all value and intensity to an entirely private domain of personal consumption'. Fisher, M. and Gilbert, J. (2013) Capitalist realism and neoliberal hegemony. *New Formations*, 80/81: p.93.

[16] Tollemache, R. (2017) Thoughts and Feelings About Climate Change: An In-Depth Investigation. Bristol, UWE. Unpublished PhD Thesis.

[17] In the UK we have come to know this as 'cakeism' – having your cake and eating it – an art extolled by the former Prime Minister, Boris Johnson.

[18] Ronson, J.(2015) *So You've Been Publicly Shamed*. London: Pan Macmillan. On-line shaming takes a variety of forms, from the targeting of specific individuals to the indiscriminate 'doxing' (the publication of private information which brings on-line anonymity to an end) of participants in on-line networks.

[19] Manago, A., Ward, M., Lemm, K., Reed, L. and Seabrook, R. (2015) Facebook involvement, objectified body consciousness, body shame, and sexual assertiveness in college men and women. *Sex Roles*, 72: 1-14.

[20] 'Objectification theory' has become one of the most powerful conceptual models in this field. Its originators Barbara Frederickson & Tomi-Ann Roberts propose that in a sexist culture, women are encouraged to evaluate themselves against internalised cultural ideals particularly regarding thinness, ideals before which they can only fail. Drawing on Helen Bloch Lewis they argue that this sense of failure leads to pervasive body shame. Frederickson, B. & Roberts, T-A. (1997) Objectification theory: Toward understanding women's lived experiences and mental health risks. *Psychology of Women Quarterly*, 21(2): 173-206.

[21] Lansky, M. (2005) The Impossibility of Forgiveness: Shame Fantasies as Instigators of Vengefulness in Euripides' Medea, *Journal of the American Psychoanalytic Association*, 53: 437-464.

[22] Surprisingly little has been written which explicitly explores the links between class and shame. Whilst shame is implicitly central to the classic study by Sennett and Cobb on the 'hidden injuries of class' the word itself is hardly mentioned. In the UK the everyday process of shaming and disrespecting through which working class children and adults are maligned and denigrated have been explored by Diane Reay and Bev Skeggs. Reay, D. (2005) Beyond consciousness? The psychic landscape of social class. *Sociology*, 39(5): 893-910. Skeggs, B. & Loveday, V. (2012) Struggles for value: Value practices, injustice, judgment, affect and the idea of class. *British Journal of Sociology*, 63(3): 472-490.

[23] Ray, S. (2021) Climate Anxiety is an Overwhelmingly White Phenomenon. *Scientific American*, March 21st, 2021. Available at:

https://www.scientificamerican.com/article/the-unbearable-whiteness-of-climate-anxiety/ (accessed 30 August 2023). However the only large-scale global survey of how young people experience climate change seems to suggest this is not a white middle class phenomenon, indeed anxiety about climate change was stronger in the Philippines and Nigeria than in European countries. See Hickman, C. et al (2021) Climate anxiety in children and young people and their beliefs about government responses to climate change: A global survey. *The Lancet Planetary Health*, December.

[24] Haslett, A. (2016) Donald Trump, Shamer in Chief. *The Nation*, 24 October, 2016. https://www.thenation.com/article/archive/donald-trump-shamer-in-chief/ (accessed 6th October 2023).

[25] McKenzie, L. (2015) *Getting By: Estates, Class and Culture in Austerity Britain*. Bristol: Policy Press.

CHAPTER EIGHT

RESSENTIMENT

Ancient wounds

I feel uncomfortable when I hear journalists and politicians referring to the war in Ukraine as the first major war on European soil since the Second World War. For the war in Bosnia seems to have been forgotten. In the early 1990s and in the heart of Europe upwards of 120,000 people, mostly Bosnians, died as a result of nationalist and racist aggression whilst 'the West' looked on and did nothing. In the last year of this war (sometimes referred to as the Balkan War at the time), I was part of a delegation to Bosnia of British support groups, trade unionists journalists and intellectuals. We were all staying in the Hotel Bristol in the war-torn city of Tuzla. After three years of war, Tuzla remained a model of a pluralist and multi-ethnic community, giving the lie to the prevailing Western narrative that this was a civil war between Serbs and Muslims.[1]

I found the group dynamics of the visiting delegation almost as thought provoking as my firsthand encounter with ethnic cleansing - the mix of Scottish trade unionists, Quaker activists and members of the intelligentsia created quite a heady brew. What follows is taken directly from my diary extracts of that time:

> Three of them are waiting for the lift in the hotel. Two of Europe's leading liberal intellectuals and a 'Yugoslav' I do not recognize. As I approach they turn their backs on me, knowing I am one of their delegation.
>
> They are discussing the Yugoslav's bright young son. Addressing the two intellectual celebrities, the Yugoslav says 'he will write for you'. One of them asks, 'Is he in Belgrade?'. 'No, he is in New York now', the Yugoslav replies. I think to myself, this is how their networks operate, through conversations in lifts where favours are traded.
>
> In the lift they pretend I am not there. The intellectuals leave on the sixth floor, the Slav on the eighth, and I am alone. These people who talk to themselves, in their capsules, protecting their 'liberalism' - protecting the right for people like them to talk to their friends.

At the time I was overwhelmed by that humiliating feeling (so familiar to members of the working class, to black people and women) of being so insignificant in someone's eyes that they don't even see you. Reflecting now I sense my own continued resentment towards this intellectual elite. How often I have ruminated on this scene: 'The shits, who did they think they were, indeed who do they think they are? How dare they do this to *me*'! And reflecting on my resentment I notice the bitterness, my injured narcissism, and the jaundiced eye I kept open for several years for any good news concerning the failing political career of one of these 'stars' as he endeavoured to lead the Liberal Party in Canada. This slight had become an ancient wound, still alive inside me as if it had happened only the day before. Now compare it to the following.

> Reporters in the Balkan wars often observed that when they were told atrocity stories, they were occasionally uncertain whether these stories had occurred yesterday or in 1941, 1841 or 1441.[2]

Here Michael Ignatieff refers to what Vamik Volkan calls a 'chosen trauma'[3] – 1441 being the year when the Serbian Prince Lazar was defeated by Murat, the Ottoman Sultan, at the Battle of Blackbirds Field. The exhumation of Lazar's remains by Slobodan Milosovic's supporters 550 years later in 1989 was the prelude to the Serb's murderous aggression towards their Bosnian neighbours. Lazar's remains were carried triumphantly around 'Yugoslav' towns and villages as support was whipped up for a resurgent Serbian nationalism fuelled by this ancient wound.

130 years ago, speaking in his essay *The Genealogy of Morals* of what he called the Master/Slave relationship, the German philosopher Friedrich Nietzsche said that whereas the noble and strong bear their suffering willingly, the rest of mankind 'excel in finding imaginary pretexts for their suffering….They revel in suspicion and gloat over imaginary injuries and slights…They tear open the most ancient of wounds'.[4] Nietzsche saw this form of resentment as part of the human condition, a kind of primordial flaw or weakness in the human character.

Nursing a grievance

Imagine a cow in a field munching away on grass. The grass is not immediately digested but fermented in a specialized stomach and then regurgitated and chewed again. This 'chewing of the cud' further breaks down the plant matter before proper digestion. This process of rechewing the cud is called rumination and animals that do this are called ruminants.

I have noticed how prone we Moderns are to ruminating and chuntering. When the modern self ruminates something subtly different occurs compared to other animals, the chewing of the cud rather than being the prelude to digestion becomes an alternative to it. Indeed, the material can often be chewed over endlessly even though this gives no nourishment, indeed quite the opposite. I wonder whether we Moderns aren't inveterate ruminators. We ruminate upon slights that we've received, gaffes we've been responsible for, opportunities we have missed, exclusions and rejections...sometimes real and sometimes imagined. Here are some examples.

Alice and a group of close friends have a WhatsApp group. One day Alice notices that several but not all group members have sent greetings to Shona on her birthday. Its late in the day, should Alice also send a greeting message? She decides not to but then spends literally days agonizing over whether Shona will have been offended and what the others in the group will be thinking about her. She becomes convinced that they think she is either self-absorbed or stupid and are talking about her behind her back. She repeatedly imagines scenarios in which she turns the tables upon the others making them feel guilty for their narrow minded judgementalism.

David is a senior doctor in the National Health Service. On Friday his new service manager emails him with a work schedule that he had not been consulted about. He is outraged. The manager is young and female, she has studs and piercings. David refers to her contemptuously as 'looking like a Tesco checkout operative'. He ruminates about this incident over the entire weekend. The weekend is thoroughly spoilt.

I could give many examples of this kind of thing from my clinical practice. These ruminations seem to share a number of dimensions – there is a strong element of complaint directed at the Other, there is a sense of being a victim (of real or imagined slights, exclusions, judgements) and, most importantly, the hurt that the self feels does not easily heal because the wound is nursed.

These are examples of what Friedrich Nietzsche called *ressentiment*, derived from the French term which literally means to 're-feel', i.e. to feel something over and over again. Now there is a short but significant slip from resentment (which English moral philosophers such as David Hume had noted was a legitimate response to injustice) to *ressentiment*. Nietzsche saw the latter as the emotion of the weak – the response of those who bite their tongues, swallow their pride and repress their anger who, resenting authority's sting, often choose to sting another, weaker figure in turn.

Nietzsche's perspective in some ways is crudely hierarchical and traditional. Crudely put he is saying that humans are pretty flawed creatures and only a tiny elite of Ubermench are able to transcend this condition and are able to bear their suffering in a noble and dignified way. The rest of us become emotional ruminants. The German philosopher Max Scheler developed Nietzsche's largely unformed concept of *ressentiment* and introduced it into mainstream philosophy.[5] He noted that for the powerless both the impulse to take angry revenge and the object of this anger are subject to repression: '[S]ince the affect cannot outwardly express itself it becomes active within. Detached from their original objects, the affects melt together in a venomous mass'.[6] *Ressentiment* then is a toxic cocktail of negative affects – bitterness, spite, envy – seeking an object.

There are many similarities between Nietzsche's idea of *ressentiment* and psychoanalytic thinking. My hunch is that when Freud spoke of the 'mental constellation of revolt' that he felt characterised the internal world of the melancholic, it was something very similar to *ressentiment* that he had in mind.[7] I also believe that John Steiner's exploration of grievance occupies the same territory.[8] We can liken *ressentiment* to the continued nursing of a grievance and in nursing this grievance the individual takes the position of the victim and in doing so enjoys the consolation of being in the right. The great thing about victimhood is the sense of one's own innocence, the responsibility for one's misfortune always lies with the Other. Hence, I (the man being ignored in the lift in Tuzla) have the consolation of being 'in the right' and this righteousness is an expression of my moral narcissism, it makes me feel good.

The agony of *ressentiment* (which on some occasions also seems to be masochistically enjoyed) is endured partly because of the consolations that moral narcissism (righteousness, being in the right, triumph, getting on one's 'high horse', etc.) brings.[9] Being in the right and avoiding being in the wrong becomes the overriding preoccupation in this state of mind. The world becomes ordered according to this calculus. Accuracy of perception and the establishment of truth become subordinated to the Law of Right or Wrong.

Beneath and beyond the everyday tribulations which provoke our *ressentiment* I believe one can find a sustained attitude of complaint towards the perceived injustices of life – the existence of minds which are different to one's own, the passing of time, the reality of the self's limitations. These 'facts of life' are also experienced as an injustice that the individual feels both victimised and aggrieved by, the grievance smouldering on (being nursed) in a melancholic fashion. In this way the fantasy of the Ideal is kept alive, albeit as paradise from which one has been unfairly excluded.

Liberal ressentiment

When you think about it, the inability to fully accept time, death, limits, loss and otherness is itself largely historically and culturally determined. For the idea that we have to live within our limits is nothing new for the global 80% who have been doing this all their lives. The paradox is that it is the 20% who are very well informed about the ecological crisis who are the least able or willing to do anything about it. Least able because we are the most hooked-in to the carbon intensive practices of globalisation (with work colleagues and loved-ones spread across the globe), and psychologically least able because our sense of narcissistic entitlement is such that when the word 'sacrifice' is mentioned we imagine that it must refer to someone else for this could not possibly apply to us. Press the point and what you get is a wave of *liberal ressentiment*.

We, the global 20%, who are disproportionately responsible for the climatic destruction presently falling upon the global poor cannot quite bring ourselves to believe that it is we who have to make some sacrifices. This indeed is 'the malady of the Ideal', of a civilization which promises what cannot be delivered. Finally freed from need via the development of capitalism's productive forces the global middle class, the new meritocracy, now falls prey to the embrace of the Ideal. Seduced, taunted and oppressed by the longing for the ideal life that it arouses we feel envious and resentful, and it is this that sews the seeds of what I now think of as *liberal ressentiment*.

In the previous chapter I noted the danger that in the liberal meritocracy, people like us, project their own sense of shame onto the white working class. I believe this dynamic can take a further step when, having constructed this stupid, unaware, narrow-minded Other we liberals then feel oppressed and persecuted by this phantom of our imagination. In the UK I believe this dynamic may partly underline some of the erroneous tactics of groups such as Just Stop Oil. When they block motorways or spoil sporting events such as snooker they justify their tactics as a way of waking people up, particularly ordinary working people, as if they, the activists, are the only ones not asleep. Facing an audience that is all too willing to assume the position of the offended victim, we step right in and assume the role of the aggressor. How tempting for us, even though we know it is counter-productive, to preach, lecture or provoke, as if we are dealing with zombies who need waking up, rather than with people who we need to engage with and listen to.

---

[1] Ahmad, J. and Hoggett, P. (1995) The death of Bosnia and the birth of the New World Disorder. *Free Associations*, 5 (3): 416-419.

[2] Ignatieff, M. (1997) The elusive goals of war trials. *Harper's*, September-October, 1997, pp. 15-18.

[3] Volkan, V. (1997*) Bloodlines: From Ethnic Pride to Ethnic Terrorism.* New York: Farrar, Straus and Giroux

[4] Nietzsche, F. (1956 [1887]) *The Birth of Tragedy and the Genealogy of Morals*. New York: Anchor/Doubleday. P.264.

[5] Scheler, M. (1992 [1912]). *Ressentiment*, Edited with an introduction by Lewis Coser, trans William Holdheim. The Free Press. Reprinted in *On Feeling, Knowing and Valuing, Selected Writings of Max Scheler* (Harold Bershady. Ed.). Chicago: Chicago University Press.

[6] Ibid, p.71.

[7] Freud, S. (1917) Mourning and Melancholia, Standard Edition of the Complete Psychological Works of Sigmund Freud, vol. 14, p.248. London: Hogarth Press.

[8] Steiner, J. (1993) *Psychic Retreats: Pathological Organizations in Psychotic, Neurotic and Borderline Patients*. London: Routledge; Steiner, J. (2013). The ideal and the real in Klein and Milton: Some observations on reading paradise lost. *Psychoanalytic Quarterly*, 82 (4): 897-923.

[9] Feldman, M. (2008). Grievance: The underlying Oedipal configuration. *International Journal Psychoanalysis*. 89(4): 743-758.

SECTION TWO

REACTIONARY STATES OF MIND AS THE HOLOCENE ENDS

For the psychoanalyst W.R. Bion,[1] the life of the mind and the life of society assumed a very similar dynamic. According to Bion the forces of truth were engaged in a revolutionary struggle against an establishment[2] which sustained its position through a barrage of propaganda which insisted that unless its position was maintained catastrophe would follow. Within the internal world of each one of us, this struggle plays itself out. Sometimes the forces of truth triumph leading to internal development, sometimes they are crushed leading to a state of paralysis in which a variety of symptoms can appear such as phobias or anxiety states. I am always reminded of the remarkably similar insight made by Antonio Gramsci the Italian Communist imprisoned by Mussolini, when speaking of periods of social crisis:

> The crisis consists precisely in the fact that the old is dying and the new cannot be born; in this interregnum a great variety of morbid symptoms appear.... The death of the old ideologies takes the form of.... politics which is... cynical in its immediate manifestation.[3]

The morbid symptoms that Gramsci spoke of – nostalgic nationalism and fascism – are obviously different from psychological symptoms but both Bion and Gramsci had struck upon the same truth in their different fields, that all forms of life manifest a dynamic balance of forces which can either lead to development or to decay. Who can doubt that today the old fossil capitalism is dying but the new is still struggling to be born. In this indeterminate position conspiracies flourish, societies fracture and strong men everywhere promise salvation.

There are perhaps four different relations to the truth. We can embrace the truth, despite its disturbing quality, and use the resulting disequilibrium to find new possibilities. We can live in fear of the truth and find ways of avoiding its disturbing consequences leading to stuckness and paralysis. We can show contempt for the truth, sneering triumphantly in its face. Finally, we may believe that we have exclusive possession of the truth. Liberal democracy lives in fear of the truth of climate change, choosing not to be too disturbed by it, ducking and weaving, saying one thing and doing another. Some of the new authoritarians show utter contempt for the truth, making up stories that they themselves don't believe in to hypnotise the 'little people'. Finally, after these snake oil salesmen have been ejected from the stage, the fundamentalists – some religious, some superficially secular – who have been lurking in the wings, make their entrance, offering salvation to all who would listen.

---

[1] Bion, W. (1970) *Attention and Interpretation*. London: Tavistock
[2] Hoggett, P. (1997) The internal establishment. In P. Bion Talamo, F. Borgogno and S. Merciai (Eds.) *Bion's Legacy to Groups*. London: Karnac Books.
[3] Gramsci, A. (1977) *The Prison Notebooks*. London: Lawrence & Wishart. p.276.

CHAPTER NINE

BUSINESS AS USUAL

'There is no alternative'

There's a phrase, much loved by parts of the left, which goes: 'its easier to imagine the end of the world than it is to imagine the end of capitalism'. In 1989 the West finally seemed to have triumphed over Soviet style Communism. Market democracy seemed to be the only show in town. Even China adopted the goal of being a 'socialist market economy' in 1992. The efficacy of the free market seemed to have been established as a self-evident truth. Margaret Thatcher referred to it as TINA – there is no alternative. It was also just at this moment of triumph that the world first began to wake up to the developing catastrophe of climate change. The Kyoto Protocol was signed in 1994 and then...nothing happened. Indeed, not only did nothing happen but more and more fossil fuels were burnt, tropical jungles were cleared, and the world's oceans were ransacked at a faster rate than ever before. It seemed that capitalism had finally triumphed, not only against its Communist opponent, but also against nature and over our collective imagination.

Mark Fisher coined the term 'capitalist realism'[1] to describe this withering of the political imagination. For Fisher this so-called realism constituted a pervasive atmosphere, a kind of invisible barrier constraining thought and action. Anything other than the status quo was simply utopian, to imagine otherwise was to be a dreamer. If this is realism then we face a very dystopian future. Indeed, some would say that we are accelerating towards the cancellation of the future.[2]

Despite the pauses caused by the global financial crisis in 2008 and the Covid epidemic in 2020, far from slowing down everything has been speeding up.[3] If cell division is the key to life then malignancy refers to the process whereby abnormal cells divide without control and begin to invade nearby tissues, as in cancer. In other words, the normal processes of life (cell division) are subject to sudden and massive acceleration and if the process goes undetected and unchecked this leads to the destruction of the organism. As the American journalist Roy Scranton asked, '(H)ow do we stop ourselves from fulfilling our fates as suicidally productive drones in a carbon-addicted hive, destroying ourselves in some kind of psychopathic colony collapse disorder?'[4]

In any animal organism there is a difference between normal healthy growth and cancerous growth. For almost three thousand years human societies have been built upon various forms of monetary exchange. But capital is a very specific form taken by money. Its only purpose is to reproduce itself in a constantly expanding fashion.[5] The role of money in an economy is similar to the role of cells in the body (each brings life to its respective system) but when, several hundred years ago, money began to assume the unique form of capital a profound shift occurred equivalent to a healthy cell mutating into a cancerous cell. If this Marxist perspective is right then capital inverts the relationship between the subject and the object, humanity becomes the slave not the master. As Margaret Thatcher also said 'you can't buck the markets'. Growth and more growth becomes the only logic of the system. Growth for what? Don't ask such a stupid question.[6]

'The Best Lack All Conviction'

Throughout the world, as soon as the Covid epidemic was over, there was huge pressure to return to 'business as usual'. In reality this outlook masks the hegemony of a business model which assumes that growth, indeed accelerating growth and accumulation, is a self-evident good, particularly if this growth has a green tinge to it. It is difficult to think of a single mainstream political party in the Global North or South which doesn't advocate growth of some kind. Successive reports by the IPCC define this 'business as usual' scenario[7] in terms of slow and incremental progress towards reaching mitigation targets, the outcome of market forces and technological innovation rather than any significant governmental intervention. IPCC's Sixth Assessment Report[8] suggest that this trajectory would lead to increases of global temperatures of approximately 2.7 degrees centigrade by 2100.

This would be catastrophic, and it remains the most likely of all the IPCC's scenarios to come to fruition given the continued absence of political will. Remember COP 21 in Paris was seen as a great advance on its predecessor COP 15 in Copenhagen and yet one is bound to wonder whether Paris was anything more than a ritualized orgy of target setting led by well-meaning liberal democracies. Soon after Paris, Angela Merkel's government fudged implementation of their 'coal exit pathway', the UK government gave the greenlight to the expansion of British airports, the $100 billion transition fund for developing countries was forgotten about and the Trump administration pulled out of the Paris Agreement in its entirety. Politicians seemed to behave as if agreeing upon the targets and signing the agreement was equivalent to 'job done'; this is an *as if* or perverse approach to change.[9] I don't have the slightest doubt that the Obamas, Merkels and now Macrons, Bidens and so on (compared to the Trumps etc.) act in good faith around climate change. But, to use a phrase from 'The Second Coming', a poem by

PARADISE LOST? The Climate Crisis and the Human Condition

W.B. Yeats,[10] the problem is that they 'lack all conviction' because they can see no alternative.

The French sociologist Bruno Latour,[11] reflecting on the arrival of the Anthropocene, noted that the vacuity of contemporary politics reflected the failure of the political class to understand 'the stunning extent to which the situation is unprecedented'. Latour was referring to the best hope we had and still have at the moment (i.e., liberal democracy), compared to the new wave of authoritarianism which is sweeping parts of the world. Unfortunately the vast majority of these progressive economic elites, the ones that gather at Davos and the UN Climate Change Conferences, still believe in 'business as usual'. They believe that with a bit of technocratic attention, some tweaking, incentivizing and stern admonishing, governments will be able to steer the market system towards a solution to the climate change crisis. Bill Gates, who seems to be a well-intentioned man, is also a living embodiment of this myopic realism. In his recent book *How to Avoid a Climate Disaster*[12] he argues that capitalism, via judicious combination of markets, government intervention, innovation and enterprise can generate the solutions required. As if more technocratic mastery, more attempts to bend nature to our will, more faith in scientific and technical progress, in short more of what has taken us to this sorry pass will now lead us out of it.

Drowning by numbers

On BBC Radio 4's early morning *Today* programme hosts can be regularly heard asking the business editor 'what are the numbers like today?'. Whilst originally a shorthand for asking about the health of the stock and bond markets 'the numbers' has become what the psychoanalyst Jacques Lacan referred to as a 'master signifier', something apparently with the power to fix meaning. We are becoming preoccupied with what 'the numbers' tell us in all aspects of life, including our personal lives, as if this is the only source of reliable knowledge.[13] Private and public corporations now live or die by 'the numbers' – targets, indicators, comparators, etc. – so much so that 'the numbers' are increasingly mistaken for the real thing. To repeat, we live in a culture today where what counts (to corporate bosses, policy makers, harassed medical professionals, Facebook obsessives and joggers) is what can be counted, and if it can't be counted then it doesn't count.

The triumph of numbers is also manifest in environmental and climate change policies underpinning the idea that we can cost the 'services'[14] provided by nature and estimate the 'discount rate'[15] to be applied to the value of future generations. I mentioned earlier in this book the language of 'mitigation banking' which argues that the depletion of natural resources renders them a commodity making them 'critical natural capital' and 'increasing their conservation value'. Mainstream economics plays a key role in climate change politics and policy making because, by putting a price on everything, a pseudo-scientific gloss can be placed upon essentially ethical and political decisions regarding the costs of climate action versus the benefits of continued inaction. This is not only about the value we give to nature but also ultimately the price of human life itself, and specifically the price we give to future generations. This requires 'putting monetary values on human suffering, harm, and even death' by attempting to estimate the 'worth' of future generations in comparison to today's. Economics professor Mariano Torras provides a lucid review[16] and critique of the shocking perversity of this approach, one exemplified by the Stern Report. The quantification of life, which is now penetrating the very way in which we think about the self (e.g. body dysmorphia, fitness addiction), lies at the heart of the controlling and disciplining impulses of neoliberal society.

And yet the impulse to quantify and measure is also essential to many sciences including climate science. Science is central to our desire to know reality and to the ethical requirement to face reality. As such, science always has a potentially subversive role in society[17] – that's why Christianity suppressed it 1600 years ago and why reactionaries dismiss it today. However science is not value free. The way in which science has developed since the rise of modernity is inextricably tied to the subjugation of nature and of other civilizations by modern Western society.[18] And this seems to be something to do with the separation of science from philosophy, ethics and the arts in the early phase of Enlightenment, a separation charted so wonderfully by Andrea Wulf's book *The Invention of Nature*.[19]

In many liberal democracies the ethos of quantification has now also taken over the very idea of what it means to govern. Governments go about setting themselves targets, developing performance measures, etc., and political debate, between the main parties at least, becomes reduced to a statistical knock about in which each side hurls its 'evidence' at the other. This is a perversion of the idea of government – quantification now becomes the means of avoiding accountability and transparency rather than assisting it.[20] When this is introduced into the realm of international governance, and specifically attempts to achieve international cooperation around climate change, then we are in big trouble.

---

[1] Fisher, M. (2009) *Capitalist Realism: Is There No Alternative?* Winchester: Zer0 Books.

[2] John Bird and Dave Green provide a valuable exploration of Fisher's work and from a psycho-social perspective, one which traces the constant interaction between the ideology of capitalist realism and the anxious and haunted inner world of the neoliberal citizen. Bird, J. and Green, D. (2020) Capitalist Realism and its Psycho-Social Dimensions. *Psychoanalysis, Culture & Society*, 25 (2): 283-293. Mark Fisher was himself a victim of capitalist realism, suffering from depression he eventually took his own life. Perhaps Mark's death was a precursor to the growing line of tragic figures who have subsequently taken their own lives having become convinced that climate change had indeed cancelled the future.

[3] A United Nations Environment Programme (UNEP) report in 2019 warned that to be on track to reach the 1.5 degrees C target emissions would need to fall 7.6% each year between 2020 and 2030. George Monbiot points out that this would be equivalent of exceeding the emissions reductions 'achieved' in 2020, the first year of the pandemic, every year for the next decade. UNEP (2019) Emissions Gap Report. Available at: https://www.unep.org/resources/emissions-gap-report-2021 (accessed 30 August 2023). Monbiot, G. (2021) Domino Theory, *The Guardian* 14 November 2021. https://www.monbiot.com/2021/11/19/domino-theory/ (accessed 30 August 2023).

[4] Scranton, R. (2015) *Learning to Die in the Anthropocene*. San Francisco, CA: City Lights Books. pp. 85-6.

[5] Rosa Luxembourg provides a classic statement on constantly expanding accumulation: 'the basic law of capitalist production is not only profit in the sense of glittering bullion, but constantly growing profit. This is where it differs from any other economic system based on exploitation. For this purpose the capitalist...uses the fruits of

exploitation not...for personal luxury, but more and more to increase exploitation itself. The largest part of the profits gained is put back into capital and used to expand production. The capital thus mounts up or, as Marx calls it, "accumulates."' Luxembourg, R. (1972) The Accumulation of Capital: An Anti-Critique. In *Rosa Luxembourg & Nikolai Bukharin: Imperialism and the Accumulation of Capital*. Edited with an Introduction by Kenneth Tarbuck. London: Allen Lane.

[6] But of course this is a vitally important question, one I will address in Chapter 23.

[7] The latest 6th Assessment Report prefers the term 'Middle of the road scenario'.

[8] The National Geographic provides an accessible summary of the likely outcomes for each of the IPCC's five trajectories or pathways. See: https://www.nationalgeographic.co.uk/environment-and-conservation/2021/08/5-possible-climate-futures-from-the-optimistic-to-the-strange (accessed 20 September 2023).

[9] Hoggett, P. (2013) Climate change in a perverse culture. In Weintrobe, S. (Ed.) *Engaging with Climate Change: Psychoanalytic and Interdisciplinary Perspectives*. London: Routledge.

[10] Yeats, W.B. (1920) The Second Coming. In *Michael Robartes and the Dancer (Poems)*. Churchtown, Dundrum: Cuala Press.

[11] Latour, B. (2018) *Down to Earth: Politics in the New Climatic Regime*. Cambridge: Polity. p.44.

[12] Gates, B. (2021) *How to Avoid a Climate Disaster: The Solutions We Have and the Breakthroughs We Need*. New York: Alfred. A. Knopf.

[13] Espeland, W. and Stevens, M. (2008) A sociology of quantification. *European Journal of Sociology*, 49 (3): 401-36.

[14] For a critical discussion of the pervasive use of the concept of 'ecosystem services' in conservation management see Bekessy, S., Runge, M., Kusmanoff, A., Keith,, D. & Wintle, B. (2018) Ask not what nature can do for you: A critique of ecosystem services as a communication strategy. *Biological Conservation*, 224: 71-4.

[15] Stern, N. (2006) *The Economics of Climate Change: The Stern Review*. Cambridge: Cambridge University Press. Commissioned by the Tony Blair government, the Stern Review was the first systematic attempt to examine the likely impact of climate change upon the world economy.

[16] Torras, M. (2016) Orthodox Economics and the Science of Climate Change. *Monthly Review*, 68 (1): 25-34.

[17] Hamilton, C. (2013) What history can teach us about climate change denial. In Weintrobe (ed.) Engaging with Climate Change: *Psychoanalytic and Interdisciplinary Perspectives*. London: Routledge.

[18] This process is described in both considerable detail and with angry lucidity in the first few chapters of T*he Nutmeg's Curse* by Amritav Ghosh. Ghosh, A. (2021) *The Nutmeg's Curse: Parables for a Planet in Crisis*. London: John Murray Publishers.

[19] Wulf, A. (2015) *The Invention of Nature: Alexander von Humboldt's New World*. New York: Knopf.

[20] Hoggett, P. (2010) Government and the perverse social defence. *British Journal of Psychotherapy*, 26(2): 202-212.

CHAPTER TEN

FEAR OF TRUTH

The era of denial

A working group to the International Geological Congress proposed 1950 as the start date for the Anthropocene[1] epoch, the first geological time to be defined by humankind's indelible footprint. Some of the most enduring aspects of this footprint comprise three toxic by-products of our recent development – radiation, carbon dioxide and synthetic polymers – the spread of which can be dated largely from the end of Second World War.

I think there's a case to be made that there is also a very distinctive toxic psychological by-product characterising this epoch. For the Anthropocene is also the epoch of denial. The problem for citizens of the Anthropocene is that we Moderns can no longer plead ignorance in the way that we could one hundred years ago. In the era of mass communications and social media it is impossible not to know about famines, mass migrations, unjust wars and, most importantly for this epoch, about ecological destruction and climate change.

These inconvenient truths threaten to overwhelm us, making normal life impossible. To manage them we create a series of splits or fractures inside ourselves between knowing and believing, and between thinking and feeling, so that what we know loses its disturbing quality. In psychoanalysis this is called disavowal.[2] It describes a particular relation to reality in which one part of the self rejects what another part of the self sees and knows. This results in a perverse relation to reality in which the truth gets twisted. It is the equivalent to what in climate change politics we call 'soft denial'[3] and it is driven by fear of the truth.

Through my own research and the research of doctoral students, and by running workshops on this issue for over a decade, I have examined in detail the many ways in which we all 'do climate denial', how we learn to stay normal in a crazy world by deceiving ourselves.[4] We say to ourselves that climate change is not an immediate danger, or it is not my responsibility, or that there is nothing I can do, or that the experts are bound to come up with a solution. These little acts of self deception help us to adjust, stave off the fear and the guilt. All of us do this in developed societies; denial becomes intrinsic to the unhappy consciousness of the Anthropocene.

Everyday disavowal

Climate change denial is part of the warp and weft of our everyday social interactions at home and work, part of what we feel we can say and do and of what we feel a pressure not to say or do.[5] It is part of the way in which well intentioned people end up staying silent[6] or not really saying what they want to say. As my colleague Ro Randall and I discovered, this can happen to climate scientists as easily as to the rest of us.[7]

Disavowal refers to that perplexing capacity we have to know something whilst not believing in what we know because what we know just seems too preposterous, too terrible, or both. So we split off our knowing from our feeling and remain undisturbed.

Make no mistake about it, the Sixth Assessment Report of the IPCC is absolutely clear. Unless we change very, very soon we're likely heading towards a 2.7 degree increase in global temperatures. The carnage is unimaginable, we're talking of hundreds of millions of dying or displaced, dwarfing the inhumanity and mortality of the two World Wars. Hannah Arendt cites David Rousset, a concentration camp survivor, when he imagined the following conversation between camp inmates:

> 'Those who haven't seen it with their own eyes can't believe it. Did you yourself, before you came here, take the rumours about the gas chambers seriously?'
> 'No' I said.
> '...You see? Well, they're all like you. The lot of them in Paris, London, New York, even at Birkenau, right outside the crematoriums...still incredulous, five minutes before they were sent down into the cellar of the crematorium...'[8]

And perhaps in a similar way I still find myself incredulous about what we're doing, as each year more and more species are wiped out, and the birds and insects disappear, and once good land becomes degraded, deserted or inundated, and more and more migrants take to the road.

The comfort blanket of everyday life

Particularly since the Second World War advances in technology have helped insulate us from wider realities, constituting a socially organised defence against troubling truths. Increasingly we Moderns live in our comfortable, protected and self-absorbed enclaves.[9] Information technology has become woven into the fabric of this infrastructure. Just-in-time systems now bring food and other essentials to our doorsteps dispensing with the need for vast back-up storage supplies. These systems and the workers that maintain them therefore constitute a kind of cocoon within which modern citizens go about their lives. We are impervious to our absolute and ruthless dependency on them, and on the nonhuman environment beyond them, until they fail us and then we feel panic and anger.

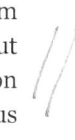

Alongside and synchronous with this infrastructure has been the development of the online world, what some call 'the virtual cocoon'.[10] Worldwide time spent online via the mobile is now estimated at an incredible 150 minutes per day,[11] dwarfing desktop internet consumption. It is almost as if we are disappearing into a virtual world, a protective bubble, a gated community in the mind. Disavowal is woven into the fabric of neoliberal society enabling us to sustain a perverse, or 'as if', relation to reality so that we can carry on with our business as usual, as if we cannot see that our house is starting to catch fire.

Our predicament

Our problem is not that we know too little but we know too much. Generations of citizens living in affluent Western democracies have learnt to live with events that are unbearable and should be intolerable. For decades we have become versed in the practice of turning a blind eye to difficult truths – genocide, mutually assured destruction and now ecocide. On each occasion we have faced the same predicament: how to normalise a reality which was not just unacceptable but which at times seemed stir crazy.[12] And this is where we are now with the climate emergency – trying to lead normal lives while all around us the conditions that first made human civilization possible 11,000 years ago are being destroyed, and by us. No wonder Amitav Ghosh calls this period 'the great derangement'.[13]

We want change *and* we want to continue with our business as usual *and* we want our leaders to tell us that we can continue with our business as usual. We want to have our cake and eat it. So, full of good intentions we will set ambitious sounding targets for 2030 and the band on the Titanic will continue to play its lullabies and we'll dress for dinner whilst not quite being able to avoid noticing the disconcerting list developing on our luxury liner.

Cultural supports for denial

There are two aspects of Western culture – entitlement and exceptionalism – which reinforce systematic denial. Sally Weintrobe has provided an incisive examination of both of them.[14] However, I believe Sally may have given insufficient emphasis to the paradoxical connection between vulnerability and entitlement. In my earlier discussion of the significance of shame in our performative culture I suggested that in neoliberal society the modern self becomes increasingly split between an entitled and abject part. Anxieties relating to personal failure have become endemic in contemporary society and in my clinical experience this includes many people who exist right at the top of the meritocratic hierarchy who are seen by virtually everyone else as a success. It is this abject part of the self that is 'hailed' by the commodity. As the advert says, 'Because you are worth it'. But the self that feels that it is 'worth it' is the exhausted, depleted, anxious self that feels entitled to some relief, to have some thing to look forward to, to be pampered, spoilt and treated as special.

The other cultural phenomenon I want to draw attention to is 'exceptionalism'. Weintrobe compares 'the exception' to the uncaring part of the self (omnipotent, invulnerable, etc) which is in conflict with the caring part. She examines some of the many ways in which exceptionalism finds expression in relation to climate change. One I would like to highlight has cropped up in many workshops I have run with ordinary members of the public. It also came up in some of the interviews with climate scientists mentioned earlier.[15] It refers to the way in which imagining oneself to be an exception leads us to assume that we are exempted from the rules which we

PARADISE LOST? The Climate Crisis and the Human Condition   83

nevertheless believe should apply to others. Whilst to some extent no doubt apocryphal we did hear stories about conservation biologists or climate change scientists scurrying to their latest important conference or field trip via combinations of taxis, intercontinental jets and 4X4's. One of the key resistances we have to changing towards low carbon lifestyles is the feeling that this really applies to somebody else and not to us, because for some compelling reason we are excused. It's as if there has been some oversight; whoever it was who compiled the checklist was clearly misinformed or got confused, because my name shouldn't be on that list as a special exemption applies to me.

---

[1] Ian Angus provides a detailed and accessible account of the emergence of the idea of the Anthropocene within the Earth Systems Science community. Angus, I. (2016) *Facing the Anthropocene: Fossil Capitalism and the Crisis of the Earth System.* New York: Monthly Review Press.

[2] I examined disavowal and provided examples of the many forms it can assume in my essay 'Climate Change: From Denialism to Nihilism' in Hollway, W., Hoggett, P., Robertson, C. and Weintrobe, S. *Climate Psychology: A Matter of Life and Death.* Bicester: Phoenix Publishing House.

[3] Hoexter, M. (2016) A Pocket Handbook of Soft Climate Denial. New Economic Perspectives, 6 October, 2016. https://neweconomicperspectives.org/2016/10/pocket-handbook-soft-climate-denial.html (accessed 30 August 2023).

[4] Several of my CPA colleagues have written insightfully about denial including Adrian Tait and Renee Lertzman. Tait, A. (2021) Climate Psychology and its Relevance to Deep Adaptation. In J. Bendell and R.Read (Eds.) *Deep Adaptation: Navigating the Realities of Climate Chaos.* Cambridge: Polity. Lertzman, R. (2015) Environmental Melancholia. New York: Routledge.

[5] Noorgard, K. (2011) *Living in Denial.* Cambridge, Mass: MIT Press.

[6] Zerubavel, E. (2007) *The Elephant in the Room: Silence and Denial in Everyday Life.* Oxford: Oxford University Press.

[7] Hoggett, P. and Randall, R. (2018) Engaging with Climate Change: Comparing the Cultures of Science and Activism. *Environmental Values*, 27: 223-243.

[8] Arendt, H. (2017 (1951)) *The Origins of Totalitarianism.* London: Penguin. p.575.

[9] Gretton, D. (2019*) I You We Them. Journeys Beyond Evil: The Desk Killers in History & Today.* London: William Heinemann. p.802.

[10] Malm, A. (2020) *The Progress of this Storm: Nature and Society in a Warming World.* London: Verso. p.12.

[11] Petrosyan, A. (2023) Daily internet usage per capita worldwide 2011-2021, by device. Statista. Available at: https://www.statista.com/statistics/319732/daily-time-spent-online-device/ (accessed 30 August 2023).

[12] In writing this book, and this chapter in particular, I realise that this issue has preoccupied me since the 1980s, indeed my 1992 book *Partisans in an Uncertain World*

opens on precisely the same issue. Hoggett, P. (1992) *Partisans in an Uncertain World: The Psychoanalysis of Engagement.* London: Free Association Books

[13] Ghosh, A. (2016) *The Great Derangement: Climate Change and the Unthinkable*. Chicago: University of Chicago Press.

[14] Weintrobe, S. (2021) *Psychological Roots of the Climate Crisis: Neoliberal Exceptionalism and the Culture of Uncare.* New York & London: Bloomsbury Academic.

[15] See Hoggett and Randall, note 7 above.

CHAPTER ELEVEN

THE INHUMAN

Writing from their firsthand experiences of the Holocaust and the terrible consequences of the Nazi invasion of Russia, Primo Levi[1] and Vasily Grossman[2] take us deep into the darker recesses of human behaviour. Their encounter with the inhumane is salutary and their conclusions are identical. Inhumanity takes two main forms: i) the dominant one is indifference and lack of feeling; ii) then there is cruelty, brutality and sadism, and the sense of triumph to be had over the abject and terrified victim.

    Their conclusions had in fact anticipated by the Irish poet and dramatist W.B. Yeats in the aftermath of the earlier world war. Yeats' poem 'The Second Coming'[3] was written in 1919 and published the following year. Many now believe that his poem is an iconic representation of the modernist imagination, presciently depicting the destructive forces gathering within contemporary Western civilization. In the poem, the gathering forces of destruction are depicted as a rough beast 'with a gaze blank and pitiless as the sun'. Years later,[4] reflecting on his work, Yeats' mentions that in 1904, in the play *On Baile's Strand,* 'I began to imagine, as always on my left side just out of range of sight, a brazen winged beast which I associated with laughing, ecstatic destruction'; he adds that this beast was 'afterwards described in my poem The Second Coming'. So it seems there were two beasts haunting Yeats' imagination, one cold and quietly murderous, the other derisive and triumphant.

'Gaze Blank and Pitiless as the Sun'

Decades ago, when women were still largely excluded from senior positions in employment, the phrase 'organisation man'[5] was used to refer to the zombie-like mindlessness of the managerial and professional staff of the modern corporations emerging in 1950s America. Trained to think but not to think too much, 'organisation man' engaged in a form of 'followership' which was eerily reminiscent of the bureaucratic desk killers of Nazi Germany. Germans after the Eichmann trial in the 1960s began to refer to the *Schreibtischtaeter,* 'the desk killer'.[6] For every Jew or Roma butchered directly at the hands of a German soldier there were literally thousands more whose death was the responsibility of the bureaucrats and specialists who designed and built the gas chambers, organised the transports, ran the factories which used the slave labour, and so on. And almost without exception this army of desk killers got away with it.

Dan Gretton has provided a shocking analysis of this disturbing phenomenon. Although he only makes a few explicit references to societal inertia in the face of ecological and climate destruction, what he does, and does so well, is to chart the blank gaze and cold heart of the modern organisation and the economic system, capitalism, which has been built upon it. Civilization and barbarism have gone hand in hand and, far from being an anomaly, the way towards the Holocaust (organised by Germany, the most 'civilized' and cultured country in Europe at the time) had been prepared via the centuries of civilizing racism and brutalizing progress that had preceded it.[7]

The key message of Gretton's book is that the businessmen, managers and technocrats who blindly collude with the malignant logic of the modern corporation are people just like you and me. As he puts it, 'something is closed down in the journey from home to work'.[8] Loving parents and considerate partners go to work and with little thought devote their working lives to organisations engaged in the practice of killing. This is what Hannah Arendt called 'the banality of evil'.[9]

So who are today's desk killers? Well there are plenty of them.[10] Gretton mentions a few, such as those that manufacture and 'pilot' the killer drones increasingly used in the conflicts in the Middle East (and now in the Ukraine). But today's most important desk killers are those that Polly Higgins[11] and others sought to bring to account by pushing for a law against Ecocide and the obvious candidates are the executives of fossil fuel companies such as Exxon, Shell, BP, HP Billiton, Anglo American and so on.

Glencore, the massive commodities trading firm established by Marc Rich in 1974, was able to happily trade with apartheid South Africa, Castro's Cuba, the Nicaraguan Sandinistas and Pinochet's Chile. According to Rich's biographer Daniel Ammann, 'he had no regrets whatsoever...He used to say "I deliver a service. People want to sell oil to me and other people want to buy oil from me. I am a businessman not a politician."'[12] In a similar vein, Gretton[13] notes that when Shell colluded with the Nigerian government in the suppression of the Ogoni people and the murder of the writer and activist Ken Saro-Wiwa, the managing director of Shell Nigeria explained '[O]ur business is to continue the business of petroleum'.

So it is the executives of such corporations that should be brought to account, and it would be easy to demonise them as somehow evil. But Gretton warns us against such 'othering' for, and this is the point he makes over again, they are just like you and me. Gretton is calling attention to what Sally Weintrobe calls the 'complex, minute, ongoing, mental acts of disassociation'[14] through which the rule of the uncaring part of the mind, organisation or society is maintained. Think about it, when we associate we make a link or connection either in the mind or in the society we are part of. When we disassociate, we do the opposite, we break a link as in disavowal where, as we have seen, the link between the thought (about climate change) and the feeling (anxiety) is broken, so that the thought remains detached, distanced and untroubling. When things get dangerous within the mind or society they are subject to what Weintrobe calls 'fracturing'.[15] We can do this individually, and we can do it in an organised way. It is in this way that organisations and their employees become blank and pitiless in their gaze.

'Laughing, Ecstatic Destruction'

But can we fully explain the collective irresponsibility involved in 'business as usual' just in terms of the survival anxieties that drive organisations to suspend their ethical obligations?[16] Or must we go further than this and consider the possibility of an anti-social, anti-life force at work? Can we fully explain the paralysis of liberal democracies and their citizens in the face of the climate emergency in terms of fear of truth and defences against anxiety or is there also some element of desire, albeit perverse desire, to be considered?

Recently[17] I concluded that what we face today in the new authoritarianism is not just denialism but an organised nihilism, a kind of insolent refusal to be worried by the devastation we are complicit in. Cara Daggett has taken this line of thinking further through her notion of petro-masculinity[18] – a proud and defiant assertion of the righteousness of fossil fuel lifestyles, exemplified by the 'coal rollers'.[19] Daggett sniffs out the specifically patriarchal character of fossil capitalism, 'the mid-20th century fantasy of American life, when white men ruled their households uncontested' which, she believes, is the hidden refrain within Trump's 'Make America Great Again'. It is this threatened masculinity which Daggett believes now finds expression in the largely symbolic gesturing of the freedom convoys and coal rollers and the desire not just to deny but to engage in a rage filled refusal of climate change.

Perhaps here we can see a self-destructive dimension, particularly in the modern masculine self. The psychoanalyst Otto Kernberg referred to self-destructiveness in this way.

> We find such self-destructive suicidal behaviour .... in severe narcissistic personalities. Here a sense of defeat, failure, humiliation, in essence, the loss of their grandiosity, may bring about not only feelings of extremely devastating shameful defeat and inferiority, but a compensatory sense of

triumph over reality by taking their own life, thus demonstrating to themselves and to the world that they are not afraid of pain and death. To the contrary, death emerges as an even elegant abandonment of a depreciated, worthless world.[20]

'Death emerges as an even elegant abandonment of a depreciated, worthless world', remember that as you think of that devastated rainforest and the uniform rows of palm oil plants that will replace it.

Credit: Pexels

A second possibility concerns envy and *ressentiment*. We Moderns were led to imagine a world without suffering, an ideal world where time was suspended and loss was eradicated. History was marching inexorably towards this point, the end of history[21] was nigh, an earthly paradise was within sight. But now reality has come along in the form of climate change and is destroying our illusion. That is so unfair! If I can't have it then no-one else shall. This is *ressentiment* aimed at life itself. The American novelist John Updike said of his central character Harry 'Rabbit' Angstrom in *Rabbit is Rich*: 'it gives him pleasure, makes Rabbit feel rich, to contemplate the world's wasting, to know the earth is mortal too'.[22] As well as an incapacity to come to terms with death as part of the natural cycle of life, Rabbit's logic is of an envious narcissism which says, 'if I have to die, then I don't want anything else to live'.

As Helen Morgan put it in some recent reflections,[23] perhaps deep within the psyche there lies an unbearable envy, a narcissistic outrage that there could be a world that will continue without me. If we are to understand and address our failure to tackle climate change fully, we need to include the idea that there is an unconscious imperative not to do so. It may even be possible that we put men in power with the necessary traits of narcissism and omnipotence that will ensure the end we also fear.

A final conjecture concerns cruelty. We have adopted a form of capitalism in which costs (such as carbon emissions, ecological destruction, etc) are construed as 'externalities'. Think of the clearance of tropical forests in Borneo by Wilmar and other corporations. The rainforest, the orangutan and the little men operating the giant chainsaws and bulldozers have no intrinsic value – it is an economic system which knows the price of everything and the value of nothing. This is cruelty masquerading as economics and neoliberalism is an extreme variant of it. When neoliberalism insists that there is no alternative, that it is the only truth, it presents itself as a form of economic fundamentalism.

So what is the relationship between fundamentalism and cruelty? Here I am reminded of Eric Brenman's essay 'Cruelty and Narrowmindedness'.[24] Brenman argues that to maintain the practice of cruelty, a singular narrow-mindedness of purpose is put into operation which has the function of squeezing out humanity and preventing human understanding in order to produce a cruelty which is inhuman. And here's the twist, he suggests that such cruelty draws its strength from love, so that in a perverse way this cruelty is practised in the name of goodness. The zealots who conceived of the holocaust and knowingly and enthusiastically implemented it did what they did out of love – out of love for their race, for their party, and for their Führer. This is the love for the group ideal that Freud had spoken of, and he noted presciently 'in the blindness of love, remorselessness is carried to the pitch of crime'.[25] Righteousness, far from only being a feature of traditional religion, is just as much a feature of secular practices such as economics, politics, science, and indeed psychoanalysis. To be in possession of the Ideal is to be in possession of that which is true and right. This is moral narcissism. Brenman notes, 'it is as though we omnipotently hijack human righteousness and conduct cruelty in the name of justice'.[26]

Just as liberal progressives are often afraid of the truth and cynics are contemptuous of the truth, there is therefore a third group who believe they are in possession of the truth. These are the fundamentalists. To be in touch with the Ideal is to be in touch with extraordinary powers: this is how fundamentalism works and has done since the time of the Crusades. For economic libertarians, freedom is such an Ideal, and it is not just democratic government that they perceive as a threat to this Ideal but the very idea of society itself.[27]

The business of cynicism

The owners and executives of fossil fuel companies have known about the long term ecological impacts of fossil fuel emissions since at least the 1970s and cynically funded disinformation campaigns just as the tobacco companies once did.[28] There was, and still is, a close relationship between fossil capital and libertarian 'thinktanks' with some executives having a foot in both camps in the USA. The Koch Brothers are the most well known, they have funded libertarian thinktanks such as Americans for Prosperity, the Cato Institute and the Heartland Institute – all of them tanks that have been designed specifically to destroy thought. The individuals behind these institutes find it hard to disguise their contempt for the little people. They have death on their hands. As Alex Steffen puts it, this group is responsible for *predatory delay*, 'the blocking or slowing of needed change, in order to make money out of unsustainable, unjust systems'.[29]

These climate denialists were not mad or stupid, far from it, they were very intelligent but also deeply calculating and cynical. I wonder sometimes whether the climate denialism of those decades became the template for the cynical relation to truth emerging in the Global North after the global financial crash in 2008. As the climate emergency deepened and international action stalled we began to see the rise of nationalist and authoritarian responses, at first prompted by the rise of climate related mass migrations. Leaders such as Putin and Trump emerged who believed in little beyond their own self-interest and the pleasure of the flex. Theirs was a deeply cynical outlook. These people peddled mistruths that they themselves did not believe in, not even the nationalist illusions that many of these malignants so freely wielded in their pursuit of power.

Contempt for the truth

Like many PCs, my computer has predictive text. A few years back when Trumpism was first gaining force in the USA I noticed that when I typed 'Trumpism' it corrected it to 'tropism'. Maybe my PC was more intelligent than I knew. Whilst tropism is commonly used to refer to the movement of plants towards sources of light it also refers to the way in which viruses/pathogens evolve to target specific hosts. If the virus can gain entry to the susceptible cell then infection will follow. One of the commonest methods of gaining entry is called endocytosis, where the virus tricks the cell into thinking that it is harmless or even nutritious. This seems to capture the relationship between the authoritarian populist and 'the people' very well.

Over 60 years ago the psychoanalyst W.R. Bion noted that our closest human relationships can take this form, one he referred to as parasitic.[30] Seductive ideas that end up sucking the life out of us Bion terms 'lies'. It is a parasitism which is spreading its viral infection rapidly across western democracies at the moment. Its carriers have included Trump, Putin, Johnson, and other populists and nationalists.

In 2017 I was listening on the radio to a Spokeswoman for Republicans Abroad. The interviewer was asking her for her thoughts on Trump's recent statement legitimising the use of torture and water boarding. The spokesman replied that of course when the Japanese had used water boarding then this was torture but when the Americans had used it then it wasn't. Feeling slightly hysterical I was suddenly reminded of a therapist colleague of mine who, upon Trump's victory in 2016, had recommended Lewis Carol's *Alice Through the Looking Glass*,[31] specifically the following encounter between Alice and Humpty Dumpty:

> 'When I use a word,' Humpty Dumpty said, in rather a scornful tone, 'it means just what I choose it to mean — neither more nor less.'
> 'The question is,' said Alice, 'whether you can make words mean so many different things.'
> 'The question is,' said Humpty Dumpty, 'which is to be master — that's all'.

---

[1] Where to start with Primo Levi, his books have affected me probably more than any other. If you haven't come across him probably the best place to start is: Levi, P. (1979) *If This Is A Man* and *The Truce*. London: Abacus.

[2] Grossman, V. (2006) *Life and Fate*, trans Robert Chandler. London: Vintage; Grossman, V. (2019) *Stalingrad*, trans Robert & Elizabeth Chandler. London: Harvill Secker.

[3] Yeats, W.B. (1920) The Second Coming. In *Michael Robartes and the Dancer (Poems)*. Churchtown, Dundrum: Cuala Press.

[4] Jeffares, N. (1968) *A Commentary on the Collected Poems of W.B. Yeats*. Stanford: Stanford University Press. P.243.

[5] Whyte, W. (1956) *The Organization Man*. New York: Simon Schuster.

[6] Gretton, D. (2019) *I You We Them. Journeys Beyond Evil: The Desk Killers in History & Today*. London: William Heinemann.

[7] Elkins, C. (2022) *Legacy of Violence: A History of the British Empire*. London: The Bodley Head; Bauman, Z. (1989) *Modernity and the Holocaust*. Cambridge: Polity.

[8] See Gretton, *I You We Them*, note 6 above, pp. 905-906.

[9] Arendt, H. (2006 [1963]) *Eichmann in Jerusalem: A Report on the Banality of Evil*. London: Penguin

[10] One of the most recent revelations concerns Facebook's suppression of its own research which indicated the harmful effect of its Instagram app on the body image of teenage girls. Gayle, D. (2021) Facebook aware of Instagram's harmful effect on teenage girls, leak reveals. *Guardian*, 14 September, 2021. Available at https://www.theguardian.com/technology/2021/sep/14/facebook-aware-instagram-harmful-effect-teenage-girls-leak-reveals (accessed 30 August 2023).

[11] Polly Higgins was a radical lawyer and environmental advocate who probably did as much as anyone to introduce the concept of Ecocide into our vocabulary.

[12] Ammann, D. (2009) *The King of Oil: The Secret Lives of Marc Rich*. London: St Martin's Press.

[13] See Gretton, *I You We Them*, note 6 above, pp.152-166.

[14] Weintrobe, S. (2021) *Psychological Roots of the Climate Crisis: Neoliberal Exceptionalism and the Culture of Uncare*. New York and London: Bloomsbury Academic., p.190

[15] Ibid. pp.225-6.

[16] Before the Nazis came to power, I.G. Farben was probably the most advanced and innovative industrial corporation in Europe. Several of its board members and many of its senior scientists and engineers were Jews, some were Nobel prize winners. For a case study of the way in which competitive pressures were used as a justification by I.G. Farben as it gradually Nazified itself see Hoggett, P. and Nestor, R. (2021) First genocide, now ecocide: An anti-life force in organisations? *Organisational and Social Dynamics*, 21(1): 97-113.

[17] Hoggett, P. (2022) Climate Change: From Denialism to Nihilism. In Hollway, W., Hoggett, P., Robertson, C. & Weintrobe, S. *Climate Psychology: A Matter of Life and Death*. Bicester: Phoenix Publishing House.

[18] Daggett, C. (2018) Petro-masculinity: Fossil Fuels and Authoritarian Desire. *Millenium: Journal of International Studies*, 47(1): 25-44; Dagget, C. (2022) Petro-masculinity and the Politics of Climate Refusal, *Autonomy*, 1 May, 2022.

[19] Rolling coal, a largely USA phenomenon, involves retrofitting a diesel truck so that its engine can be flooded with excess gas, producing thick plumes of black smoke.

[20] Kernberg, O. (2009) The concept of the death drive: A clinical perspective. *International Journal Psychoanalysis*, 90(5): p. 101.

[21] For Francis Fukuyama the collapse of the Soviet Empire by 1991 announced the final triumph of liberal democracy. The American form of governance and society henceforth represented the end-point towards which all other societies would inexorably gravitate. Fukuyama, F. (1992) *The End of History and the Last Man*. New York: Free Press.

[22] I am indebted to my colleague Wendy Hollway for putting me in touch with this line from Updike. Updike, J. (1981) *Rabbit is Rich*. New York: Knopf.

[23] Personal correspondence.

[24] Brenman, E. (2006) Cruelty and Narrowmindedness. In Brenman, E. and Spoto, G.F. *Recovery of the Lost Good Object*. London: Routledge.

[25] Freud, S. (1921) Group Psychology and the Analysis of the Ego, Standard Edition of the Complete Psychological Works of Sigmund Freud, Vol. 18, p.113. London: Hogarth Press.

[26] See Brenman, Cruelty and Narrowmindedness, note 24 above, p.257.

[27] Brown, W. (2019) *In the Ruins of Neoliberalism: The Rise of Antidemocratic Politics in the West*. New York: Columbia University Press.

[28] Oreskes, N. and Conway, E. (2010) *Merchants of Doubt: How a Handful of Scientists Obscured the Truth on Issues from Tobacco Smoke to Global Warming*. New York and London: Bloomsbury.

[29] Steffen, A. (2016) Predatory Delay and the Rights of Future Generations, *Medium*, 29 April, 2016. Available at https://medium.com/@AlexSteffen/predatory-delay-and-the-rights-of-future-generations-69b06094a16 (accessed 22 September 2023).

[30] Bion, W. (1970) *Attention and Interpretation*. London: Tavistock.

[31] Carol, L. (1871) *Alice Through the Looking Glass*. London: Macmillan.

EXCURSION

THE SS ARMENIAN

Trevose Head lies on the Atlantic coast just west of Padstow in north Cornwall. In terms of its ruggedness, it is not as wild as the coast near Hartland, my 'go to' place. That being so, the last time I visited Trevose I was amazed by the number of skylarks almost invisible as they hung above us singing their wonderful song. To the east of Padstow, just across Daymer Bay, lies the village of St. Minver. In St. Menefreda Churchyard there is a memorial dedicated to unknown SS Armenian Seaman. The SS Armenian was thought to have sunk out in the Atlantic off Trevose Head on June 28th 1915. It was transporting fourteen hundred mules from Newport News, Virginia, in the USA to Bristol to replace horses that had died in and around the trenches during the First World War. The SS Armenian was intercepted by a German submarine. When it eventually surrendered the German commander gave the remaining crew the option of taking the two remaining lifeboats and striking for the Cornish coast (a Belgian trawler rescued the survivors the following day). The Armenian was then torpedoed and all the mules were lost.

But this is by no means the full story. Twelve of the crew, having developed a sincere affection for their mules,[1] declined the offer of the lifeboats and chose to sink beneath the waves whilst tending the ship's living cargo. Little is known about the life of the muleteer on these Atlantic crossings but something of the flavour of the conditions during the war can be gleaned from the statement of J.M. Garret, a white muleteer on a vessel called the Nicosian.

> Each muleteer was required to attend from twelve to twenty mules. They were confined to their stalls and were not cleaned until they were discharged. The sleeping quarters were separated by a plank partition and the odours were disgusting.
>
> The sleeping accommodations were a shaving mattress on a plank bunk, with one woollen blanket and a shaving pillow. The fare consisted of Irish potatoes, spoiled meat, and pea soup. Coffee in the morning and tea at night, and one small bun twice a day, one in the morning and one at night.[2]

The twelve muleteers on the Armenian who chose to die with their animal comrades were African American muleteers, part of a larger group of 77, and they refused to abandon the mules for whom they had developed sincere affection and respect. No doubt there was also a solidarity of the oppressed at work here. From what I have been able to ascertain from existing records[3] the twelve men were:

> John Frobey
> Le Roy Harris
> Julius Henry
> Robert Jackson
> Thomas King
> Andrew Little
> Walter Oakes
> Henry Ryckcerd
> Howard Small
> John Smith
> Edgar Wall
> William Young

We should honour these men for the choice that they made. Somewhere out there, perhaps 24 miles west of Trevose Head in what is called the Celtic Trench, their bodies still lie.

[1] Kalaydjian, T. (2013) Built During the Hamidian Massacres, Sunk During the Genocide. The Armenian Mirror-Spectator, March 30th, 2013. Available at http://www.mirrorspectator.com/2013/03/20/built-during-the-hamidian-massacres-sunk-during-the-genocide/ (accessed 6 October 2023).

[2] Statement on Oath by the United States Citizen J.M. Garrett, of Kiln, in the County of Hancock. Available at http://www.vlib.us/wwi/resources/archives/texts/t050925/Baralong.html (accessed 6 October 2023).

[3] SS Armenian (+1915) Available at: https://www.wrecksite.eu/wreck.aspx?11109 (accessed 6 October 2023).

CHAPTER TWELVE

THE ARMED LIFEBOAT

The new authoritarianism

Although Nietzsche, Scheler and Steiner see *ressentiment* primarily as an aspect of the human condition, Scheler also brought a sociological perspective to the understanding of the powerlessness that feeds *ressentiment*. Social *ressentiment* is fuelled in societies which profess equality but in practice sustain massive inequality. If in addition there are no political parties or movements capable of giving voice to this injustice then a palpable sense that 'no-one speaks for us' develops and the ensuing feeling of political impotence provides the condition in which grievances become suppressed and turn in upon themselves. These are absolutely crucial insights (with enormous contemporary relevance) and one reason why Scheler was so widely referred to by political theorists who studied the populist and authoritarian movements and parties which emerged throughout the Western world in the 1920s and 30s.[1]

Writing during the Reagan/Bush years, Christopher Lasch[2] railed at an American left that he felt had abandoned a programme of political and economic reform in favour of what he saw as a self-indulgent agenda of cultural and moral radicalism. Lasch traced the origins of the US 'culture wars' in the late 1960s to an 'unspoken agreement not to raise questions about the distribution of wealth' by both Republicans and Democrats which meant that social and cultural rather than economic issues dominated election campaigns from the 1980s onwards.[3] Later the American philosopher Richard Rorty[4] developed the same criticism of the left's embrace of cultural politics, believing it would presage the return of political reaction built upon a wave of white working class resentment.

Lasch argued that a new global elite was emerging out of the growing knowledge, financial, and cultural industries.[5] This elite was both economically and socially liberal, pro-globalization, multiculturalist, meritocratic, devout in its worship of progress, and with no loyalty to place. It was also an elite which, he argued, was more completely out of touch with the conditions of growing inequality facing the working class and lower middle class than any equivalent elite since the inception of American society.

Lasch would have been mortified to discover that Steve Bannon, a key Trump advisor and one of the main orchestrators of the political reaction that he feared, has been strongly influenced by his writings. Bannon's white,

Christian nationalism, now being replicated in Putin's Russia, Orban's Hungary and Meloni's Italy, offers an imaginary Ideal, that of an exceptional or chosen nation or people[6] which could be made great again. In his musings on the reactionary mind, Mark Lilla notes the apocalyptic inflection that nostalgia can take both in white nationalism and fundamentalist Islam. He adds perceptively, 'for the apocalyptic imagination, the present, not the past, is a foreign country'.[7]

Climate change, nationalism and authoritarianism

Influenced by writers such as Jem Bendell,[8] several leading figures within Extinction Rebellion (XR) in the UK have warned of the danger of civilizational and social collapse[9] in the wake of the climate emergency. For my own part I believe that a lurch into authoritarian nationalism is a much more likely initial response to the deepening climate crisis in the coming decades. This is what we should be frightened of, not catastrophist fantasies of panicked crowds fighting in the aisles of supermarkets as the shelves run dry. The first signs of this imminent authoritarianism could be discerned in the rise of populist and nationalist movements in Western Europe in the first decade of the new millennium. Figures such as Viktor Orban (Hungary), Marine le Pen (France) and Geert Wilders (Netherlands) anticipated Putin's increasingly authoritarian turn and, later, along came figures such as Trump and Bolsanaro.

The nationalist alternative is effectively based upon the assumption that that the game is already up, that attempts to mitigate climate change are either dangerous or futile, and that the political task is to provide the conditions for societal survival in the face of the environmental and social chaos to come. To achieve this it seeks to decouple capitalism from globalization. This is a bit like a whole nation preparing to hunker down to pass out the coming storm, erecting their own 'gated communities' no longer in Palo Alto but across whole parts of the globe. Following Christian Parenti this has become known as 'the politics of the armed lifeboat'.[10]

The idea of a 'lifeboat ethic' was first propounded by the influential Texan ecologist Garrett Hardin in 1974.[11] Hardin invited the reader to imagine a rich nation as a lifeboat in an ocean full of more numerous poor people who would dearly like to climb aboard. Hardin argued that it would be wrong to help the poor in this way because 'the boat swamps, everyone drowns. Complete justice, complete catastrophe'.[12] For Hardin 'survival' was the criterion with which the consequences of such actions should be measured. Hardin's work has been very influential within environmentalism, particularly his idea of the *tragedy of the commons,*[13] that is, the idea that resources

PARADISE LOST? The Climate Crisis and the Human Condition 99

that we hold in common such as the oceans will ultimately become depleted by individuals acting rationally in their own short-term interests. Hardin's stance, informed by Darwinist and Malthusian assumptions, generally led him to quite reactionary conclusions regarding immigration, population control, food aid and other policies. Not surprisingly, Hardin's ideas had some influence over some of the coming generation of climate change scientists and policy makers, and this can be seen clearly in the work of James Lovelock.[14]

Half a century later and Hardin's lifeboat scenario has literally come to life as wave upon wave of citizens from the global South, uprooted by drought, flood, civil conflict and collapsed states, seek safety in Europe and North America. This has spurred the development of new forms of authoritarian nationalism, preoccupied with securing the nation's borders against the waves of migrants. In its more extreme variants, it has fuelled paranoid political fantasies such as the theory of 'The Great Replacement', which imagines a white community becoming outnumbered by non-white, particularly Muslim, incomers.

Photo courtesy of Owen Newman

Securing borders, energy and food

In his excellent book *Facing the Anthropocene,*[15] Ian Angus notes that the politics of the armed lifeboat was anticipated in a report commissioned by the Pentagon as far back as 2003.[16] This nationalist response with its emphasis on food and energy security and the protection of borders[17] is anticipated in one of the IPCC's five future possible scenarios in its Sixth Assessment report. This scenario reads:

> **Regional Rivalry – A Rocky Road (High challenges to mitigation and adaptation)**
>
> A resurgent nationalism, concerns about competitiveness and security, and regional conflicts push countries to increasingly focus on domestic or, at most, regional issues. Policies shift over time to become increasingly oriented toward national and regional security issues. Countries focus on achieving energy and food security goals within their own regions at the expense of broader-based development. Investments in education and technological development decline. Economic development is slow, consumption is material-intensive, and inequalities persist or worsen over time. Population growth is low in industrialized and high in developing countries. A low international priority for addressing environmental concerns leads to strong environmental degradation in some regions.

The IPCC's Sixth Assessment Report estimates that this scenario would probably lead to global temperature increases of 3.6 degrees,[18] that is, an even worse prospect than 'business as usual'.

Suddenly, in the wake of Putin's war on Ukraine, the phrase 'energy security' is upon every politician's lips. No longer an abstract idea talked about by academics, each nation begins to focus upon how it can secure its own energy free from dependency upon others. And as climate chaos and political instability intensifies, the same is already beginning to happen with 'food security'. The politics of the armed lifeboat is a perfectly realistic and all too probable strategy for managed social collapse where, at least for a while, the social fabric of some societies (mostly in the Global North) is maintained whilst others disintegrate under the escalating pressures of heat, drought, famine and war. Seen from this perspective the new breed of nativist political leaders – the Trumps, Putins, Orbans, Le Pens and so on – are the would-be commanders of the Titanic's lifeboats, but these are lifeboats which are armed to the teeth and busily kicking away the drowning survivors of the catastrophe.

[1] In his introduction to the first English translation in 1961 of Scheler's book *Ressentiment*, Lewis Coser notes its impact on the generation of scholars writing in the immediate aftermath of the rise of Nazism and the Second World War. Scheler, M. (1912 (1992)). *Ressentiment,* Edited with an introduction by Lewis Coser, trans William Holdheim. The Free Press. Reprinted in *On Feeling, Knowing and Valuing, Selected Writings of Max Scheler* (Harold Bershady. Ed.). Chicago: Chicago University Press.
[2] Lasch, C. (1991) *The True and Only Heaven: Progress and its Critics*. New York: W.W. Norton.
[3] Ibid, p. 508.
[4] Rorty, R. (1999) *Achieving our Country: Leftist Thought in Twentieth-Century America.* Cambridge: Harvard University Press.
[5] Lasch, C. (1996) The Revolt of the Elites and the Betrayal of Democracy. New York: W.W. Norton.
[6] Smith, A. (2003) *Chosen Peoples: Sacred Sources of National Identity*. Oxford: Oxford University Press.
[7] Lilla, M. (2016). *The Shipwrecked Mind: On Political Reaction.* New York: Review Books. P.137.
[8] In their introduction to a recent edited collection, Jem Bendell and Rupert Read speak of 'societal breakdown or collapse' and add that by this they mean 'an uneven ending of industrial consumer modes of sustenance, shelter, health, security, pleasure, identity and meaning'. Bendell, J. and Read, R. (2021) (Eds.) *Deep Adaptation: Navigating the Realities of Climate Chaos*. Cambridge: Polity, p.2.
[9] The theme of collapse is an aspect of what, in Chapter 14, I call the 'apocalyptic imagination'. This state of mind is even more familiar to the far right than to the left. Leading thinkers on the right in the USA have been anticipating such collapse since at least the 1980. They include the libertarian John Pugsley, predictors of nuclear war such as Bruce Clayton, evangelical Christians such as Jim McKeever and economists such as Barton Biggs.
[10] Parenti, C. (2011). *Tropic of Chaos: Climate Change and the New Geography of Violence*. New York: Nation Books. My colleague Sally Weintrobe refers to this politics as Noah's Arkism Twenty-First Century Style. Weintrobe sees a powerful connection to exceptionalism, as she puts it, 'I am entitled to be saved and entitled to have the world as I want it saved for me. I, uniquely and personally, and my group, will face no real loss'. Weintrobe, see note 21, p.252.
[11] Hardin, G. (1974) Lifeboat ethics: The case against helping the poor. *Psychology Today*, 8: 38-43.
[12] Ibid, p.38.
[13] Hardin, G.(1968) The tragedy of the commons. *Science* 162: 1243–1248.
[14] Hoggett, P. (2011) Climate change and the apocalyptic imagination, *Psychoanalysis, Culture and Society*, 16 (3): 261-275.
[15] Angus, I. (2016) *Facing the Anthropocene: Fossil Capitalism and the Crisis of the Earth System.* New York: Monthly Review Press.
[16] Schwartz, P. & Randall, D. (2003) An Abrupt Climate Change Scenario and its Implications for United States National Security. Washington DC: US Department of Defence. Available at https://training.fema.gov/hiedu/docs/crr/catastrophe%20readiness%20and%20response%20-%20appendix%202%20-%20abrupt%20climate%20change.pdf (accessed 4 September 2023).

[17] The control if not elimination of immigration was a fundamental element of reactionary nationalism even before climate change came along. With the climate crisis the protection of national borders from migrants fleeing drought, famine and war becomes central to the politics of the armed lifeboat. On the relation between climate change and immigration I recommend the work of the Climate and Migration Coalition. https://climatemigration.org.uk/ (accessed 4 September 2023).

[18] Stone, M. (2021) 5 possible climate futures – from the optimistic to the strange. *National Geographic*, 20th August, 2021. Available at: https://www.nationalgeographic.co.uk/environment-and-conservation/2021/08/5-possible-climate-futures-from-the-optimistic-to-the-strange (accessed 4 September 2023).

CHAPTER THIRTEEN

REACTIONARY STATES OF MIND

In Chapter 8 I argued that *ressentiment* was the core sentiment underlying populism and authoritarianism. *Ressentiment* is the sense of grievance that one's Ideal (one's God, community, way of life or country) is at risk or has already been damaged or stolen by the Other.[1] As we have seen, for many European populists this Other is Islam, and its threat is represented in the fantasy of the Great Replacement.[2] In contrast, political reaction in the USA and Brazil seems to revolve more closely around the perceived threat to white masculinity presented by feminism, multi-culturalism and globalisation.[3] Perhaps then it is no coincidence that the slogan 'God, Country, Family' is one that reactionary movements on both sides of the Atlantic can unite around in their quest to restore an imagined Ideal of the past.

Whilst the Trump's and Putin's of this world are contemptuous of the truth and have a cynical relationship to the populist beliefs that they peddle, this is certainly not the case with other reactionary figures such as Le Pen or Bolsanaro who appear to be true believers. Nor is this the case for many members of nationalist movements for whom the slogan 'God, Country and Family' is full of meaning, the total truth. So does this mean that such nationalism is a form of proto-fascism? Or is there is a vital difference between the nationalist fervour of the new authoritarians and more fundamentalist political movements which imagine themselves to be the bringers of an earthly paradise?

Authoritarianism and totalitarianism

In her study of totalitarianism Hannah Arendt repeatedly makes the point, fascism was not in the business of creating another authoritarian dictatorship in Europe, it sought much more than this.[4] As the Yale historian Timothy Snyder[5] points out, what inspired the Nazis was the idea of *lebensraum* or 'living space', that is, the creation of an Aryan super state occupying most of Europe (including much of Russia). The Nazis saw the Aryan race as the apotheosis of an imaginary biological-historical struggle.[6] Nor was Bolshevism nationalistic (unlike Stalinism). For pure communists, history was the history of the class struggle, and communism was 'understood to be the most progressive and highly developed stage of history'.[7] Both ideologies were therefore utopian. Although religion might promise salvation in the afterlife, these secular systems of thought sought heaven on earth, in the material reality of a perfected society.

In contrast to the incoherent, often contradictory, programmes of populist and authoritarian leaders, totalitarian movements are generated and sustained by rigid but coherent ideologies. It follows that whilst authoritarian regimes such as Putin's Russia are typically imposed from above, totalitarian regimes such as Khomeini's Iran were originally generated from below.

Everything connects

Arendt offers an incisive analysis of totalitarian ideology. It is characterised by a form of thinking that claims to be so in touch with the truth as to be unperturbed by the disconfirmations of messy reality; that is, it is in touch with a supposedly truer reality 'concealed behind all perceptible things, dominating them from this place of concealment and requiring a sixth sense that enables us to become aware of it'.[8] Arendt adds that it is this sixth sense that one can acquire by becoming an adept of the ideology; by mastering the important texts, one is able to intuit the hidden meanings and secret intents behind public events (this way conspiracies lie). Subscribers to totalitarian systems of thought believe that they are in possession of the total truth, hence the term totalitarian.

Traces of the totalitarian state of mind can be discerned in populism and authoritarianism. The spread of conspiracy theories became a striking feature of the Trump presidency[9] but they really began to take hold of the imagination of multifarious groups of Western citizens during the Covid pandemic. The beliefs involved conspiracies about the nature and spread of Covid (including the idea that it was a hoax and didn't actually exist, or that it did exist and was deliberately introduced by the Chinese/Bill Gates/the deep state/5G), about anti Covid vaccines, about the wearing of masks, the 'real' reason for lockdowns, and so on.

What was striking was the susceptibility of the alternative health/deep green/wellbeing movement to many of these theories, a phenomenon given the name 'conspirituality' by Jules Evans.[10] Remember, one of the three distinguishing characteristics of totalitarian thought was, according to Hannah Arendt, its claim to be able to discern hidden truths 'requiring a sixth sense' for us to become aware of them. According to Arendt, 'it always strives to inject a secret meaning into every public, tangible event and to suspect a secret intent behind every public political act'.[11] As many have commented when friends or acquaintances have become drawn to conspiracy theories, it all happens so suddenly. One moment they're talking about a particular yoga exercise they have found enriching or expressing interest in a Zach Bush or Charles Eisenstein podcast, and then next moment they're avidly following cult alt-health guru Kelly Brogan. Brogan is what today is called a 'social influencer'. Through her website, podcasts, subscription based chat rooms, and so on, she has acquired a small but significant influence, particularly among middle class women in the USA. Championing 'natural' health as opposed to 'medicalised' health care, Brogan originally drew support from women opposed to a medical establishment which was seen as corporate and patriarchal. Medical expertise and authority was not to be trusted.

Then along came Covid and it was but a short slip from a largely harmless and slightly New Age critique of medical authority to an increasingly pernicious criticism of the public health profession in the USA and its representatives such as Dr Anthony Fauci. A criticism which started with scepticism about the purported virulence and contagiousness of the virus, quickly spread to opposition to lockdown, mask wearing and other preventative measures, and then on to championing full blown anti-vax conspiracies involving the Gates Foundation, the use of nanotechnologies in the human bloodstream, etc.

In a fascinating article, internet game designer Reed Berkowitz takes the reader through the perverse logics of QAnon and other conspiracy generators.[12] Berkowitz introduces a new term, *apophenia*. As the tendency to perceive a connection or meaningful pattern between unrelated or random things, *apophenia* appears to be one of the key psychological processes underlying the attraction of conspiracy theories.

There is virtually nothing in the psychoanalytic literature on apophenia. The term was developed by the German psychiatrist Klaus Conrad who saw it as a crucial feature of the early stage of schizophrenia. To cite Conrad,

> Borrowing from ancient Greek, the artificial term 'apophany' describes this process of repetitively and monotonously experiencing abnormal meanings in the entire surrounding experiential field, eg, being observed, spoken about, the object of eavesdropping, followed by strangers.[13]

Conrad notes that the grip of these meanings on the patient's mind is so powerful that they are unable to transcend their current experience or shift their frame of reference. Returning to conspirituality, whilst those gripped by conspiracies such as The Great Reset might speak in terms of having an epiphany, the reality is that they are in the grip of an apophany, a kind of semantic promiscuity where hidden meanings lurk everywhere.

Everything connects, hence the seductiveness of this style of thinking to the alt-health and deep green milleux with its emphasis on holism and integration. Jules Evans reminds us how rife this convergence of New Age-ism, occultism and fascism was during the Nazi era.[14] Evans notes the absorption of astrology, parapsychology, alternative medicine, 'back to nature' ecocentrism and nature-mysticism and organic farming into the Nazi worldview alongside conspiracy theories regarding the grip of a global Jewish cabal on the world economy, media and government.

A future totalitarianism?

Towards the end of her analysis of totalitarianism in power, Arendt insists, that 'radical evil has emerged in connection with a system in which all men have become equally superfluous'.[15] This theme of superfluousness was one that Arendt returned to again and again, particularly in her analysis of the concentration camps. She had already warned us that it was the experience of a savage and pointless war followed by mass unemployment and social dislocation in the 1920s that sowed the seeds of this mentality. She concludes: '[T]otalitarian solutions may well survive the fall of totalitarian regimes in the form of strong temptations which will come up whenever it seems impossible to alleviate political, social or economic misery in a manner worthy of man'.[16] My belief is that the conditions fostering the social chaos that Arendt feared could lead to a resurgence of totalitarianism are already returning. We are facing an epistemological crisis, a crisis of knowing, one that Arendt likened to the pre-Totalitarian period when 'everything is possible and nothing is true'.

The deepening climate crisis is already prompting the emergence of a new authoritarianism. This is the real danger we currently face, not some kind of hypothesised collapse into anarchy. Many within these new authoritarian movements really do believe in conspiracies such as the Great Reset and in the absolute truth of slogans such as God, Country and Family. In other words there is a fundamentalist strain within authoritarian movements. Followers believe in the absolute rightness of their way of thinking, they believe they are in possession of the fundamental truth, of the Ideal. But, unlike Nazism and Communism, presently the truths that are upheld are backward looking and nostalgic. They seek to restore an Ideal – an exceptional nation, white Christian values, etc. – which is thought to be in imminent danger. In this respect the authoritarian and nationalist movements we are seeing today are, so far, significantly different to the totalitarian mass movements we associate with Nazism and Communism. These were forward[17] looking movements which sought to create an earthly paradise in the very near future.

So what might a totalitarianism of the Anthropocene era look like? Well my guess is that it would be Utopian, very modern and centred upon a fervent belief in an abstract Ideal. It would be much more likely to emerge from Silicon Valley or the plains of Nevada than the Republican heartlands of the Midwest or the industrial wastelands of Ohio. Perhaps something like a mutant strain of transhumanism, one which construed both nature and humanity as essentially superfluous?

---

[1] For the idea of 'stolen enjoyment' see Zizek, S. (1990) Eastern Europe's Republics of Gilead. *New Left Review*, 183: 60-61.

[2] At first the theory of the Great Replacement was promulgated by far-right white nationalists who believed that the original Christian majority of their nation's population was in imminent danger of being replaced by Muslim and therefore Islamicised incomers. More recently the theory has come more simply to refer to the replacement of 'indigenous' whites by non-whites.

[3] As I mentioned in Chapter 5, Wendy Brown sees the rise of political reaction in the USA very much in terms of the struggle to preserve white male supremacy in the face of the modernising force of neoliberalism. In a previous chapter I mentioned that in a similar vein Cathy Daggett has developed the idea of 'petro-masculinity' to understand the attraction between political reaction and fossil fuel capitalism.

[4] Arendt describes the regimes that came to power in Italy, Spain, Portugal and many Central and Eastern European countries before the Second World War as 'nontotalitarian dictatorships'. She argues that Rohm and the SS Stormtroopers had to be liquidated by Hitler in the 'night of the long knives' precisely because they sought to impose a military dictatorship in Germany. Arendt, H. (2017 [1951]) *The Origins of Totalitarianism*. London: Penguin. pp.404, 476-7, 484-5.

[5] Snyder, T. (2015) *Black Earth: The Holocaust as History and Warning*. London: Bodley Head.

[6] Overy, R. (2004) *The Dictators: Hitler's Germany, Stalin's Russia*. London: Bodley Head. pp.267–269.

[7] Ibid, p. 267

[8] See Arendt, note 4 above, p.618.

[9] The most prominent of these at first concerned the existence of a cabal of child sex traffickers involving prominent Democrat politicians operating out of a pizza restaurant. Later, conspiracies about election fraud and of the election having been stolen began to flourish.

[10] Evans, J. (2020) 'Conspirituality' – the overlap between the New Age and conspiracy beliefs. *Medium*, 17 April 17, 2020. Available at: https://julesevans.medium.com/conspirituality-the-overlap-between-the-new-age-and-conspiracy-beliefs-c0305eb92185 (accessed 4 September 2023).

[11] See Arendt, note 4 above, p.618.

[12] Berkowitz (2020) A game designer's analysis of QAnon. *Medium*, 30 September 2020. Available at: https://medium.com/curiouserinstitute/a-game-designers-analysis-of-qanon-580972548be5
(accessed 4 September 2023).
[13] Conrad K. (1959) Gestaltanalyse und Daseinsanalytik. *Nervenarzt*, 30: 405–410.
[14] Evans, J. (2020) Nazi hippies: When the New Age and the far-right overlap. *Medium*, 4 September 2020. Available at: https://gen.medium.com/nazi-hippies-when-the-new-age-and-far-right-overlap-d1a6ddcd7be4 (accessed 4 September 2023).
[15] See Arendt, note 4 above, p.602.
[16] Ibid, p.603.
[17] I hope it is clear that I am not using 'forward looking' in the moral sense but more in the sense of whether it is the past that is yearned for or the future.

SECTION THREE

FACING DIFFICULT TRUTHS

One often hears that phrase, taken from T.S. Eliot, that humankind cannot bear too much reality. I wonder about which kind of human is being thought about here. Is it those, like me, who live in warm houses, who never have to worry about where their next meal is coming from, who never have to deal with daily humiliations handed to them because of the colour of their skin or the 'uncultivated' way in which they speak? It strikes me that most of humanity has a lot of reality to bear, and it is we protected members of the modern meritocracy, who have least to bear, who seem the most ill equipped to handle it.

As we have seen, denial is probably the most common way of not bearing reality and disavowal is the most common form of denial. In fact it is so common that it has become our normality, for to disavow is to know something without really believing it, thereby remaining undisturbed. In a moving account of the life and death of his mother Susan Sontag, David Rieff recounts how, even though cancer had been a recurring feature of Sontag's life for 30 years, she refused to accept the possibility of her own mortality. Rieff notes the paradox that 'so terrified of death she could not bear to speak of it, my mother was also obsessed with it'. She constantly visited cemeteries, named her second novel *The Death Kit* and was drawn to places of death such as Sarajevo in the early 1990s. Using the example of his dying mother Rieff notes that there are moments when reality simply breaks through whatever filters, narratives or propaganda we surround ourselves with. Two weeks before she died, she noted:

> I was in her hospital room in Seattle when, months after the transplant, when she could not roll over in bed unassisted and was hooked up to 300 metres of tubes infusing the chemicals that were keeping her alive but could do nothing to improve her condition, her doctors came in to tell her that the transplant had failed and the leukaemia was now full-blown. She screamed out in surprise and terror. 'But this means I'm dying,' she kept saying, flailing her emaciated, abraded arms and pounding the mattress.[1]

The parallel with climate change is striking. Like sea defences, there comes a point in time when the social defences against anxiety that have been built up to keep out reality are breached, and the climate and ecological crisis begins to pour through in a disturbing way. In the Global North this finally began to happen on a large scale round about 2019. It was then that the phenomenon of what became known as eco-anxiety began to appear in the media.

Many of us in the climate psychology movement prefer to use the term eco-distress rather than the more journalistic eco-anxiety. As we shall see there are two fundamentally different forms of this distress depending upon whether it is fuelled by a preoccupation with the survival of the self or a preoccupation with the survival of the other, particularly the other (human and more than human) upon whom we depend. The first manifests in acute anxiety, the second in despair and grief. The first can lead to a paranoid and survivalist state of mind, the second either to despair or to the mobilisation of care and action. These two reactions emanate from two different parts of the self which can be found in all of us Moderns.

So in this section I explore the nature of eco-distress; how can we contain it, transform it and use it to sustain an engagement rather than disengagement with the deepening climate crisis? If we come out of denial, do we inevitably fall into despair, panic and flight, or can the truth paradoxically bring relief and strengthen our ethical resolve? In other words, if we are to fully face up to the desperate situation we are now in, that our house really is on fire, might this galvanize us rather than paralyse us?

---

[1] David Rieff (2008). Why I had to lie to my dying mother, *The Guardian*, 18 May 2008. Available at: https://www.theguardian.com/books/2008/may/18/society (accessed 21 September 2023).

CHAPTER FOURTEEN

FEELINGS OF PANIC AND FANTASIES OF ESCAPE

Catastrophic anxiety

Humanity has become a force of nature, hence the use of the term the Anthropocene to denote the era we are now in. But even before the great acceleration of carbon emissions we had, through the invention and then proliferation of nuclear arms, become a force of nature. I am old enough to remember this period and particularly the height of the Cold War in the period following the Iranian Hostage Crisis in 1979, when it really did feel to me and many of my generation like the end was nigh. At the time I had two sons, one a baby and one a toddler. We lived in Brixton in South London. At times during this period I felt utter panic. I was woken repeatedly by nuclear nightmares, bombs literally going off inside me. I went back into psychotherapy. Like millions of others I fantasised about escape even though one part of me recognised that in the case of Mutually Assured Destruction there was nowhere to escape to. Then, as now, many of us dreamt of moving to New Zealand; the best I could manage was to move my family a hundred miles down the road to Bristol. Joel Kovel, a psychoanalyst who was soon to give it up and move into environmental and other forms of activism, got it just right when he referred to this period as 'the state of nuclear terror'.[1] The world had got itself into a terrible state and our minds had followed down the same path.

In my clinical experience of working with people experiencing acute forms of climate distress there is often an overwhelming sense of imminent catastrophe which triggers 'end of the world' feelings combined with the sense that one is completely and utterly alone. This can be accompanied by a sense of shock and disorientation as our taken for granted world (the absolutisms of everyday life), with its certainties and routines disappears. Loss of appetite, sleep difficulties, constant rumination and panic attacks may ensue.

So the danger is that we switch from denial to terror, from complacency to catastrophe and that 'end of the world' feeling. I believe that it is this anxiety which now increasingly characterises our collective psychology. It is becoming a structure of feeling[2] for the end times provoking a variety of forms of survivalism. It connects to something in all of us. The survivalist in us fearfully imagines a coming social collapse, perhaps even the end of civilization.[3] Perhaps, as I suggested in chapter 3, the catastrophe that drives survivalism is ultimately one inside us, a basic fear that lurks within all of us individualised Moderns.

A basic fear?

When we are in a state of terror we are overwhelmed by an unrepresentable fear to which the only reactions are freeze, panic, fight and flight.[4] Thinking of the experience of catastrophe, one of the physical symptoms of the more severe cases of the coronavirus has been a lack of oxygen. Some hospital patients have been interviewed and have given a graphic depiction of the effect of this on them. They say that you feel that you are gasping for air, that there is just not enough getting into your lungs, it makes you feel very panicky. By all accounts it's a terrible feeling, presumably that's why torturers conjure up a similar effect through waterboarding.

These thoughts take me back to the now rather neglected psychoanalyst Michael Balint. Balint insists that in the prenatal and immediately postnatal environment there are no hard or sharp objects, but rather a 'harmonious interpenetrating mix-up'.[5] He then offers two metaphors of such interpenetration, the fish in the sea and the air that a person breathes. In each case we have a medium that the fish/person takes totally for granted. But, as with the coronavirus patient, deprive them of this medium and their response is vehement – it feels like a catastrophe to which they respond violently.

Now admittedly this is conjecture, but I wonder if the experience of terror isn't in some way connected to a rupture in a medium in which we are all immersed but which, like oxygen, we take totally for granted until we are suddenly deprived of it. Balint argues that to become human we have to learn to swim in a new medium, no longer the amniotic sea but a human medium. This calls attention to a basic trust involved when, being with others, we immerse ourselves in this human medium and in doing so establish contact with the emotional life of the group.[6]

We begin life held within an amniotic sea, immersed in the nonhuman environment of organic nature. To become human we have to learn to swim in a new medium, a primary social medium, before death brings our sojourn within the human to what seems to most Moderns like a hasty end. And I wonder whether this journey isn't haunted by the fear of catastrophe, that at any moment we might drown, suffocate or fall though space and that as a result some of us never accomplish this task fully and never feel fully able to trust this social medium. Without its reliable presence our life is precarious, at any moment like a fish out of water or a person without air we can feel that

PARADISE LOST? The Climate Crisis and the Human Condition   113

our world is ending. Perhaps this is what Bion had in mind when he coined the term 'catastrophic anxiety'.[7]

These psychoanalysts clearly believe that their conjectures apply to all human beings, that the human psyche is structured around a basic fault or flaw which generates a primitive form of anxiety. Later experiences of shock, loss, violence, disaster, etc. trigger this anxiety and escalate the fear making it more difficult to contain and manage. The thing about anxiety, that makes it so difficult for all of us to deal with it, is that at first it doesn't have an object. It is 'free floating' and haunts us, it is visceral, spreading across the chest and shoulders like a film of petrol on water. So in key move we externalise it, 'I fear' becomes 'I am frightened of' – it is much easier to deal with an external threat than an internal sense of endangerment. Now we can mobilise our aggression against this danger. So a state of mind emerges which is paranoid in its orientation.

As should now be clear, I no longer hold to the idea that all human minds are structured in this way. Both psychoanalysis and psychology are the products of modernity and the individual they describe is actually the Modern self and not some human universal. Remember the fundamental markers of this self, its separation from the other, from nature (i.e. from the more than human) and from its own creaturely nature. These separations constitute the basic fault within the psyches of we Moderns.

Apocalyptic survivalism

Catastrophic anxiety provides the conditions for survivalist ideologies, groups and practices to flourish. Today survivalism assumes so many forms. It can be religious or secular, hi or low tech, it can involve an imagined escape to the hills, to New Zealand, to Mars or heaven or to a retreat inside the self.[8]

Workers and citizens become preoccupied with their survival at any cost. Survivalism infects the culture of the working environment which takes on the hue of living in an extreme situation – narrow your horizons, focus on what is immediately before you, travel lightly without attachments, be vigilant, take on the protective colouring of your surroundings, and so on.[9]

It's not that survivalists fear a coming catastrophe, the point is that they are already living the catastrophe. The end times have already arrived and conspiracy theories which reveal their secrets proliferate like Arachne's web. Catastrophic anxiety has become a cultural phenomenon, a structure of feeling of our age.

Christopher Lasch examined survivalist fantasies at the time when terror of Mutually Assured Destruction (MAD) was at its height in the early 1980s. He noted that alongside ordinary individualised survivalism there was a particular group fantasy he called the 'apocalyptic imagination'. According to Lasch, 'the apocalyptic vision appears in its purest form not in the contention that the nuclear arms race or uninhibited technological development might lead to the end of the world but in the contention that a saving remnant will survive the end of the world and build a better one'.[10] The saved constitute an elite of the clear sighted who have the foresight to prepare for the worst and the moral fibre to prevail.[11] There is what Lasch calls a 'pseudo-realism'[12] about this way of thinking, one which is contemptuous of what it sees as the myopia of those (in peace or environmental movements) who believe that it is never too late to act. Such beliefs are seen as consolatory illusions, the resort of the weak. Tough mindedness means being able to think the unthinkable.

Christopher Lasch wrote his book *The Minimal Self* after the Iranian Hostage crisis and during the Reagan/Thatcher years. His thesis was that at an unconscious level people were already preparing for the worst, 'sometimes by building fallout shelters and laying in provisions, more commonly by executing a kind of emotional retreat from the long-term commitments that presuppose a stable, secure and orderly world'.[13] His book examined the ordinary and not so ordinary forms that this survivalism was taking, one of which was the retreat into the self. My fear is that parts of the green milieu are being infected by this.

Green survivalism

In the early 1980s, when I wasn't frozen in terror at the prospect of nuclear annihilation, I imagined some kind of geographical flight. As I said earlier, I didn't quite manage New Zealand though I did get as far as Bristol. But there are other forms of flight, other kinds of destinations offering other forms of salvation, some of which can have a corrosive effect upon the struggle to mitigate climate change, especially when they are adopted by those who purport to champion an eco-centric ethics.

If an iconoclast is a person who attacks settled beliefs or institutions, then Paul Kingsnorth's relation to environmentalism has been that of the iconoclast. Having been an environmental direct activist in the 1990s by 2010 he had, in a style similar to Lovelock, begun to deride attempts to develop renewable energy solutions such as wind and solar. Then he set up the Dark Mountain project, whose pessimism regarding the imminent collapse of civilization preceded networks like Deep Adaptation by almost a decade. In recent years he's come out as a supporter of Brexit, and now as a convert to Eastern Orthodox Christianity. He's certainly not someone who goes with the crowd!

Where Kingsnorth will end up is anyone's guess, so it is interesting to follow his recent trajectory.

A recent review of Kingsnorth's three novels – *The Wake*, *Beast* and *Alexandria* – concludes, 'Kingsnorth's true allegiance…. is to the Earth that humanity appears intent on destroying'.[14] This is not an abstract Earth, the one that moves astronauts when they look down upon it, but a literal earth, the earth of soil, land, place, folk and kin. For Kingsnorth this is undoubtedly central to his personal sense of identity and meaning, and he assumes this is the case for all of us.

Writing soon after Trump's ascendency to the White House, Kingsnorth reflected on the rise of populism across Europe and the USA.[15] I couldn't help noticing that whilst he initially appears to fault neoliberalism for the rise of populism, this mutates into a focus on globalism, which he sees as a dangerous cultural project: 'Traditions, distinctive cultures, national identities, religious strictures, social mores – all would dissolve away in the healing light of free trade and a western liberal conception of social progress'. The end of his article returns to the theme of his novels when he says: 'Our oldest identity is one that still holds us in its grip, whether we know it or not…we are all animals in place'. And it is this identity which he feels is threatened by globalism.

But his critique of globalism has mixed up two very different phenomena – the bland commercial uniformity of Apple and Big Mac with the (multi) cultural creativity and heterogeneity of the typical European city. Both equally products of globalism but as different as chalk and cheese. And by mixing them up this way one has to conclude that he is antipathetic to both of them, and sees them as somehow a threat to 'our oldest identity'.

So it seemed that Kingsnorth was becoming a cultural conservative. But what he seemed to want to conserve was not the 'British way of life' that still means so much to many Conservative Party supporters (a 'way of life' incidentally which is still inextricably connected to the myth of Empire) but some much more archaic idea of identity. As if we were to travel back in time far enough we would get back in touch with our own indigenousness. Indeed, the struggles of indigenous peoples around the world appears to be one of the few struggles that Kingsnorth now feels able to identify with.

And now in a further twist to the Kingsnorth trajectory, in what he calls his 'Vaccine Moment'[16] he says that his 'intuition screams to us that something is wrong'. And it turns out that what is wrong is that a new, radical form of techno-authoritarianism is unfolding before our eyes. What really threatens us now, he believes, is not Covid, and not climate change (which he seems to rarely mention these days), but 'the vast grid of technological control that I call the Machine'.[17]

I have long felt that in his deep antipathy to civilization (what he now calls 'the Machine') and in establishing the Dark Mountain project he epitomised an apocalyptic state of mind. So I couldn't help feeling a bit smug when in this latest post of his, Kingsnorth proclaims: 'We live in an apocalyptic time, in the original sense of the Greek word *apokalypsis*: revelation'. But we should know by now where the politics of revelation takes us, we have seen enough of it in the last four years in the USA - the politics of hidden forces, secret cabals and conspiracies. It takes us to some very dangerous places.

Back in 2017 Kingsnorth seemed to be advocating what he called 'a benign ecological nationalism'.[18] But as the current trickle of climate refugees fleeing to Europe and the UK turns into a flood I fear where any kind of nationalist solutions will lead us. There are no national solutions to climate change. Kingsnorth must know that. Perhaps that is why he no longer talks about climate change because for him the game is already up and the time has come to hunker down in his rural spiritual retreat.

Love divided

When our defences crumble and waves of climate distress crash in upon the psyche, we have seen that one of our reactions is to become preoccupied with the self and its survival. To be clear, when I refer to the self here I'm not just referring to the individual self but equally to the 'we' with whom the self identifies narcissistically. This 'we' thinks as one, it is an imagined community which can take the form of a religious movement, a nation or a race. It *is* love that binds such groups, but it is a narcissistic love, one where we look into the face of another group member and see ourselves. Within this imagined community, care, compassion, respect for difference and kindness can flourish so long as it is to one of us. There is a particularism about this kind of love, one that coexists easily with indifference or outright hostility to the stranger.

---

[1] Kovel, J. (1983) *Against the State of Nuclear Terror*. London: Pan.

[2] Drawing on the work of Raymond Williams, sociologists have used the term 'structure of feeling' to refer to feeling states shared by groups, communities, sometimes whole societies, that endure over time. Williams, R. (1977) *Marxism and Literature*. Cambridge: Cambridge University Press.

[3] There is a crucial difference between survivalist fantasies of the end of civilization and those who speak in terms of the end of civilization as we know it. Whereas the former is consumed by anticipation of the end of the world the latter is glimpsing the possibility of new beginnings.

[4] Many years ago I wondered whether there is something indwelling to our psyches which could be likened to a basic fear. Melanie Klein appears to suggest that this triggers a foundational psychic reaction through which 'I fear' becomes 'I am frightened of' and the (unknowable) danger within becomes the danger without against which aggressive defences can be organised. In this way 'man' doth go looking for monsters to slay. For Klein this unknowable danger within is what Freud referred to as the death instinct, a kind of anti-life force in the mind and society. Hoggett, P. (1992) A place for experience: A psychoanalytic perspective on boundary, identity and culture. *Environment and Planning D: Society and Space*, 10: 345-356.

[5] Balint, M. (1968) *The Basic Fault: Therapeutic Aspects of Regression*. London: Tavistock Publications. p.66.

[6] Bion's early work on the dynamics of the unstructured groups that he conducted at the Tavistock Centre in the 1940s convinced him that the task of establishing contact with the emotional life of the group was a formidable one, and that the achievement was always to some extent precarious. Bion, W. (1961) *Experiences in Groups*. London: Tavistock Publications. p. 141.

[7] Meltzer, D. (1978) *The Kleinian Development: The Clinical Significance of the Work of Bion*. Perthshire: Clunie Press. pp. 110-118.

[8] Bradley Garrett offers a detailed analysis of the preppers subculture and the various fortified retreats and refuges they are busily constructing. Garrett, B. (2020) *Bunker: Building for the End Times*. London: Allen Lane

[9] Cohen, S. and Taylor, L. (1992) *Escape Attempts: The Theory and Practice of Resistance to Everyday Life*. London: Routledge.

[10] Lasch, C. (1984) *The Minimal Self: Psychic Survival in Troubled Times*. New York: W.W. Norton & Co, p.83.
[11] Ibid, p.81.
[12] Ibid, p.86.
[13] Ibid, p.16.
[14] Wagner, E. (2021) Paul Kingsnorth and the new climate fiction. *New Statesman*, 17 March 2021.
[15] Kingsnorth, P. (2017) The lie of the land: Does environmentalism have a future in the age of Trump. *Guardian*, 18 March 2017. Available at: https://www.theguardian.com/books/2017/mar/18/the-new-lie-of-the-land-what-future-for-environmentalism-in-the-age-of-trump (accessed 21 September 2023). For a strong critique of this position see, Out of the Woods (2017) Lies of the land: against and beyond Paul Kingsnorth's volkisch environmentalism. Libcom.org, 31 March 2017. Available at: https://libcom.org/article/lies-land-against-and-beyond-paul-kingsnorths-volkisch-environmentalism (accessed 21 September 2023).
[16] Kingsnorth, P. (2022) The Vaccine Moment: Covid, Control and the Machine. Available at: https://www.paulkingsnorth.net/vaccine (accessed 4 September 2023).
[17] Ibid.
[18] Kingsnorth, The lie of the land, see note 15 above.

CHAPTER FIFTEEN

GRIEF, MOURNING AND DESPAIR

Following the publication in 2018 of the IPCC Special Report on Global Warming of 1.5°C[1] and the heatwaves and bushfires of 2019, climate anxiety began to gather, manifest in public meetings and on social media, and percolated into the psychotherapy consulting room. It's more than thirty years since the influential psychoanalyst Hanna Segal wrote her paper on the threat of nuclear war.[2] Rereading it I notice both similarities and differences to the predicament we are now in. The same mechanisms of denial and disavowal in relation to the danger are to the fore. But the threat then was one of instant annihilation, probably of all of humanity, whereas now the danger creeps insidiously but relentlessly upon us, and upon some more than others. Back then Segal felt that our own destructive impulses were denied and projected into the other group, the Russians, against whose hostile intent we sought an imaginary deterrence. Now, as we systematically vandalise the living systems upon which we all, humans and nonhumans, depend, there is no enemy 'other' to blame. Our destructiveness is exposed starkly before us. It would be tempting to speak of 'species shame' and 'species guilt' if only it weren't for the inconvenient fact that those of us (white, middle class, Western) who were and still are most responsible for this mess are those who, to begin with, will be least affected. It follows that one of the big challenges for people like 'us' coming out of disavowal is how to manage the moral emotions – grief, guilt and shame - which focus upon our concern for both the human and nonhuman other and which, like terror, can feel quite overwhelming.

Grief ongoing

When Caroline Lucas, Britain's only Green MP, announced her intention to retire one of the things she planned to do in the future was to be a doula. The role of the doula is to be a companion providing support at critical moments in a person's life. A doula is normally involved in the process of giving birth, but in Caroline's case she planned to train as a doula for the dying. In an interview she said, '(I)ts about dealing with grief and understanding different forms of grief. It made me think about what so many people are feeling now in terms of the planetary emergency: the preciousness, the fragility'.[3]

To those who glibly say, 'don't mourn, organise', Caroline Lucas' repost was to say we must mourn and we must organise, for grief and grieving lie at the heart of climate action. To be truly climate aware is to become a companion to the dying.

Just the other side of the ridge, which was the topic of my first Excursion, there is a giant playing field, about half a mile in length. In the summer I loved to stand there with my dog as swallows swooped and manoeuvred inches above the ground, at times almost within arm's reach. Once, during the COVID pandemic, I watched as a woman stood in the middle of the field laughing in utter astonishment at their aerobatics. I returned to the field recently, the swallows had all gone. Their absence was palpable, the keening made all the more piercing for knowing that in a decade's time very few locals will know of the wonder that could once be found here every summer.

Unlike the death of a loved one, species loss is not a single event after which some kind of return to normal occurs, for this is a dying without end. First the swallows, then the starlings, then what? We are involuntary companions to this dying,[4] as the Sixth Great Extinction gets underway. Our wounds will stay open, our grief will continue.

In the mid 1970s, as the Galtieri dictatorship got underway in Argentina, approaching 30,000 mostly young activists were 'disappeared' by fascist death squads and the security services, the two working hand in hand. Mothers who were looking for their children began assembling in a square opposite the Presidential Palace and as their numbers grew and they became known for their brave, dignified, silent protest. They took on the name of the Mothers of the Disappeared.[5] A group of these mothers, led by Hebe de Bonafini, continued their protest for 30 years until they felt that a full accounting for the actions of the perpetrators had been achieved. Throughout these decades they remained a persistent force pushing for democracy and social justice in Argentina. In their newspaper *Madres* they once stated, 'Let there be no healing of wounds. Let them remain open. Because if the wounds still bleed, there will be no forgetting and our strength will continue to grow.'

The Mothers made a deliberate choice to keep their wounds open. In contrast we have no choice as the dying continues all around us. We have to learn to bear the tragedy of it, not broken by it but certainly broken open. This requires something akin to what Susan Kassouf calls a 'traumatic sensibility'.[6]

On despair

Despair is common to all of us in our private lives. It is what we feel when these moral emotions threaten to overwhelm us, when the damage that we face in our personal lives seems beyond repair. Frances was one such person.

Her father was a deeply damaged man who, growing up during the war, had not known that the young woman he thought was his older sister was in fact his mother, a family secret which had been hidden from him and the rest of the world. Years later he became an angry and controlling husband and father. As a child Frances was terrified of him, but also loved him and would do anything to try and please him. In later life Frances chose a husband whose own childhood had been ruined by two alcoholic parents and sadly he became the new focus of her lifelong project to make the men around her happy, to repair damage which was actually irreparable. When she came to see me Frances was in a state of deep despair. She was wracked by guilt and persecuted by the feeling that she was a disappointment to her husband who she felt she let down continually. She could see no way out. The despair had almost immobilised her. Drained of energy, in her therapy sessions even talking exhausted her.

The experience of Frances illuminates two important facets of despair. We may be faced with an 'object'[7] of our love which seems beyond repair. Alternatively, we may feel that whilst what we love is not necessarily beyond repair, we nevertheless lack the inner resources which would enable us to carry out the repair that is needed.[8]

Applying this now to the environment, the despairing part of us believes that the destruction presently being visited upon nature is irreversible. When I wait anxiously for the return of swifts to my skyline each year my fear is that one year they will not return and worse, that this will be a portent of something worse to come, that the annual migration of swifts from Equatorial and Southern Africa to Europe will cease altogether. That wonderful species of swift will be no more.[9] In my despair its imminent extinction takes on the status of a certainty, the forces stacked against its continued survival seem too great and my fear for them overwhelms me. Reflecting on this for a moment, in reality I cannot know whether the extinction of the swift is inevitable, it may be but it may not. But the despairing part of me cannot deal with uncertainty, in a perverse sort of way it takes comfort from the certainty of despair rather than endure the excruciating precarity of hope's tightrope.

So what about the other facet of despair, that we may simply lack the inner resources to repair the damage? For Frances this was manifest in her tragic belief that she was failing her husband and therefore that she deserved his criticism and must try harder. And the more she tried the more depleted and exhausted she became and the more he criticised her. Does this apply to the environmental movement? Yes of course, we call it burnout. Burnt out activists not only feel that same depletion and exhaustion but also that same tyrannical internal critic telling them they must do more.[10]

From resignation to doomism

Whereas terror leads to panic or flight, despair can lead to resignation, fatalism and the abandonment of hope. I can feel that in myself sometimes. That feeling that it is just too difficult, that the forces resisting change are just too extensive and well resourced, that our own efforts are just too puny, a drop in the ocean. On such occasions my anger will still be there, but it is a kind of resigned anger, wrapped in a deep sigh. My motivation ebbs. Internal propaganda starts up: 'Paul, you're becoming an old man, you deserve to enjoy the time you have left…how deluded are you to imagine you can make any kind of difference to this parlous state we're in?'

Of course it's not far from here to the writer Jonathan Franzen's conclusion regarding his own feelings of powerlessness regarding climate change, 'what makes intuitive moral sense is to live the life I was given, be a good citizen, be kind to people near me, and conserve as well as I reasonably can'.[11] As the eminent climate scientist Michael Mann and others have noted there is a deep fatalism running through Franzen's outlook. Although he remains a staunch 'birder' and conservationist he encourages 'inactivism' on climate change, the biggest issue facing us.

The accusation of encouraging inactivism is the charge made by Mann against a range of figures in the climate movement.[12] Mann criticises climate scientists like Kevin Anderson and Will Steffen, journalists like David Wallace-Wells[13] and activists such as Rupert Read, as well as more obvious doomsters[14] such as Guy McPherson[15] and Roy Scranton.[16] Mann's argument is that by giving too much emphasis to the risks and dangers that face us these figures encourage fatalism and inactivity.

Despite the enormous consensus about the basics of climate change within the scientific community, considerable areas of uncertainty nevertheless remain. These include the nature and imminence of positive feedback loops such as methane release (and thus the likelihood of non-incremental change) and the actual likely rate of increase of global temperatures given the enormous range of physical, economic and political factors still in play. When Ro Randall and I interviewed a small group of leading climate scientists in the UK we were struck by the considerable pressure scientists faced from policy makers to collude with a cautiously optimistic perspective and not say anything which might be construed as 'alarmist'.[17]

That said the question remains, just how do we deal with radical uncertainty in our everyday lives as well as in politics and society? Nearly twenty years ago now, in an analysis of Tony Blair's decision to collaborate in the invasion of Iraq, I noted that where facts are not black and white but filled with caveats, qualifications and possibilities, the capacity of the group to manage the grey area of uncertainty becomes paramount.[18] On the one hand lies the Scylla of complacency (a perverse refusal to see what is in front of you), whilst on the other lies the Charybdis of paranoia (a tendency to read 'too much' into what is present).[19]

Climate realism

Unlike denialists, realists recognize that the danger of climate change is real, that the impacts are already occurring, and that they are affecting some more than others. They also recognise that surface temperatures would already be much greater if it wasn't for the vast amount of additional heat now being stored in the world's oceans, and that in time this stored heat will begin to affect surface temperatures on land.[20] In other words, even if decisive policy and market change was now occurring (and it isn't), things would continue to deteriorate for quite some time. Our trajectory is downwards. This is not doomism it is simply acknowledging what the science tells us. Doomism is to say there is nothing to be done, we're heading for the end of all human and most nonhuman life on earth.[21] I don't believe that. And yet at times I do catch myself fearing that in full knowledge of our folly, no longer in denial but in despair perhaps, or in hysteria or enraged, we will spend our last decades as a condemned man might, waiting for the scaffold.

So I believe we have a responsibility to think the unthinkable and then speak the unspeakable. In a sense this is what psychotherapists try to facilitate in their practice, to be able to help the other face the difficult truths that at that moment they are too fearful to face. There *is* still time left to act. We can't limit the rise in global temperatures to 1.5 degrees, and the odds against 2 degrees lengthen with every day of global inaction. But the fact remains that the sooner we can bring emissions under control, the better the chance we give ourselves of avoiding crash landing into the Anthropocene. As Bill McKibben recently put it: '[I]f we miss the two-degree target, we will fight to prevent a rise of three degrees, and then four. It's a long escalator down to Hell.'[22]

Many climate scientists do now concur with Bill McKibben's view – that the direction of travel is now downwards, that the final stop isn't worth thinking about, but there are many stops on the way. Jane Goodall, the primatologist, put it in a slightly more positive way, that there's still an awful lot left and that's what we've got to fight for. So the question is how do we galvanise and mobilise publics for the least worst scenario? As the saying goes, to try and to fail or to only partially succeed is not to be defeated – to be defeated is never to try.

[1] IPCC (2018) Special Report on Global Warming of 1.5 degrees C. Available at https://www.ipcc.ch/sr15/download/ (accessed 4th September 2023)

[2] Segal, H. ((1987) 'Silence is the real crime', *International Review Psycho-Anal*, 14: 3-12.

[3] Williams, Z. (2023) 'It's lonely in parliament': Caroline Lucas on life as Green MP – and what she'll do next. The Guardian, 9th June, 2023. Available at: https://www.theguardian.com/politics/2023/jun/09/parliament-always-felt-like-a-nightmare-caroline-lucas-on-life-as-a-green-mp-and-what-shell-do-next (accessed 24 September 2023)

[4] And there's something about the imperilled swifts and swallows which is all the more poignant for the way in which they represent for me and many others another group of imperilled migrants, but this time human, as more and more will flee from the climate death zones of Africa and Asia.

[5] I provided a case study of the Mothers of the Disappeared in Hoggett, P. (2009) *Politics, Identity and Emotion*. Boulder, Col: Paradigm. Pp.18-22.

[6] Kassouf, S. (2022) Thinking Catastrophic Thoughts: A Traumatised Sensibility on a Hotter Planet. *American Journal of Psychoanalysis*, 82: 60-79.

[7] The language of 'objects' and 'object relations' has dominated much psychoanalytic thinking since the Second World War. Important figures in our lives do not find literal representation in our inner world but are complex amalgams of real experience and phantasy – i.e. internal objects. The self engages in complex relationships with these objects and sometimes, as with Frances, these internal relationships are repeated over and over again in our mind, thereby contributing to the formation of our character.

[8] The Kleinian current within psychoanalysis sees such despair arising when we are overwhelmed by what it calls 'depressive anxiety'. Segal, H. (1975) *Introduction to the Work of Melanie Klein*. London: Hogarth Press.

[9] The migratory swift that comes to the UK is but one of the 113 species of swift inhabiting the temperate and tropical regions of the earth. For me its loss is no less difficult to bear knowing that it is but one of many.

[10] Hoggett, P. and Randall, R. (2018) Engaging with Climate Change: Comparing the Cultures of Science and Activism. *Environmental Values*, 27: 223-243.

[11] Franzen, J. (2018) *The End of the End of the Earth*. London: 4th Estate. p.51.

[12] Mann, M. (2021) *The New Climate War: The Fight to Take Back Our Planet*. London: Scribe. Chapter 8.

[13] Wallace-Wells, D.(2019) *Uninhabitable Earth: Life After Warming.* New York: Random House.

[14] Andreas Malm offers a pretty sharp and to my mind persuasive criticism of Scranton and Frantzen (albeit from a rationalist perspective) in the last chapter of *How to Blow Up a Pipeline*. Malm, A. (2021) *How to Blow Up a Pipeline*. London: Verso.

[15] McPherson, G. (2019) Becoming hope-free: Parallels between death of individuals and extinction of homo sapiens. *Clinical Psychology Forum*, 317: 8-11.

[16] Scranton, R. (2018) *We're Doomed. Now What?* New York: Soho Press.

[17] See Hoggett and Randall, note 10 above. I notice that Mann uses the pejorative term 'aggressive' to describe the views of Steffen and Anderson, neither of whom have been willing to collude with this pressure from policy makers.

[18] Hoggett, P. (2005) Iraq: Blair's Mission Impossible. *British Journal of Politics and International Relations*. 7: 418-428.

[19] The phenomenon I described in terms of apophenia in Chapter 13.

[20] Climate science indicates that approximately 0.5°C is currently 'baked in' to the earth and the oceans.

[21] And I agree that this is the position of figures such as Guy McPherson, but is this true of Roy Scranton? When Scranton says 'we're all in Portbou with Benjamin' I think he speaks eloquently of the despair we all face. The question is, what do we do in the face of this despair? Very honestly, but perhaps a bit flippantly, Scranton tells us that at that point, standing above the memorial near where the writer Walter Benjamin took his own life in the Spanish border town as he abandoned his struggle to escape Nazism, Scranton's thoughts kept turning to lunch and he had a sandwich. And this is the criticism to be levelled at Scranton, when push comes to shove he appears to offer little beyond irony as he pauses and meditates upon our collective folly. See Scranton, We're Doomed, note 16 above, pp.74-6.

[22] McKibben, B. (2018) How Extreme Weather is Shrinking the Planet. *New Yorker*, November 26, 2018. Available at: https://www.newyorker.com/magazine/2018/11/26/how-extreme-weather-is-shrinking-the-planet (accessed 21 September 2023).

EXCURSION

SCARP

Photo courtesy of Owen Newman

Travel west and slightly north from Tarbert, the main harbour on Harris in the Outer Hebrides, and you eventually come to the long single track road, the B887 to Hushinish. In the right weather this is a most extraordinarily beautiful road, to the south the sea loch Loch a'Siar and the island of Taransay and on a clear day glimpses of Luskentyre beach. To the north the mountains stretch away to Lewis, perhaps a hundred square miles and now completely uninhabited.

As a twenty-year-old on my first visit to Harris I remember hitching a lift from an eccentric young upper class English couple and their son. I'd heard that there was a kind of hostel on a small island called Scarp just a few hundred metres across the sound from Hushinish. I took food with me, for although there were crofts on Scarp there were no such things as shops for miles around. The ferry across to Scarp was a rowing boat with an outboard motor and from what I remember it was operated by a boy in his early teens. The hostel was a bothy with peat stove and tilly lamp. You let yourself in. It was empty. I think I was there three days and didn't talk to a soul. I remember one day walking the circumference of the island in the mist and rain, accompanied all the way by a local border collie.

PARADISE LOST? The Climate Crisis and the Human Condition

I revisited the Outer Hebrides a few times in my early twenties but didn't go back to the island. It was not until the occasion of my seventieth birthday that I managed to get back to Harris again. I wanted to show my partner some of its wonders – the Machair, the beach at Luskentyre, the vast stretch of sand beyond Scarista, and of course the island of Scarp. We drove out to Hushinish, walked the half mile to the Sound of Scarp and gazed across at the now deserted island.

I had already understood that Scarp was no longer inhabited but to my surprise I discovered that Scarp's last two remaining families had abandoned the island in December 1971, just four months after my earlier visit. The Scotsman spoke of one of the families 'leaving on a small boat packed with furniture as their two cows swam behind them'.[1]

Intrigued, I started to do some research on what had happened and this opened up new perspectives regarding Scarp's past and present.[2] In the early nineteenth century there had been eight families on the island living on its leeward side looking across to the mainland. They lived off a few cultivated plots and what the sea had to offer. But when the Highland Clearances occurred Scarp suddenly became a refuge for many victims. Some were tenant families who were evicted from more fertile ground on the mainland opposite by the new laird, Alexandra Macrae. Others came from farther afield as the following testimony, collected by local historian Donald J Macleod, indicates:

> Allan MacLean from Scarp who was a teacher had a terrible hatred of landowners and Tories. He told me many horrendous stories of the suffering of the people including his own. They were evicted from Ardroil (Capadal), Uig, trekked through the mountains and eventually settled amongst the bogs I think at Torray or another village. Then they were evicted from there and had to load their boat with all their family, chattels etc and sail for Scarp where they settled.[3]

Here is another example picked up in my own research:

> Flora Maclennan (1766-1853) originated from Scarp. Flora married Murdo Macinnes from Luachair; the couple settled at Luachair and had five children. The family moved to Brollum in Southern Pairc and when they were cleared from there, moved to 13 Old Orinsay. Orinsay was cleared in 1843 and they went to live with their son Alexander at 5 Crossbost, this being their fourth move when the couple were coming up to their 80th year.

Far from being an armed lifeboat the Scarpachs offered sanctuary to the victims of the clearances. As a consequence of such multiple evictions Scarp's population for much of the nineteenth century was beyond what was sustainable, peaking at 213 in 1881. More than 25 families eked out an existence on Scarp until after World War One. Although a few Scarpach families were eventually given crofts on the Harris mainland many were forced to emigrate, along with other Hebrideans, to eastern Quebec. Gisla Cemetery in Quebec is the final resting place of one hundred and seventy of these Hebrideans:

> *This small cemetery is located on Gisla Road near Milan, Quebec. From Route 214 heading east take Gisla Road about 3 miles. Caution: this gravel road winds through a forest, and is very narrow. Suddenly the cemetery will appear in a clearing on the right. If you visit here in the summer, be wary of mosquitos and black flies.*[4]

Evidently many of those who first landed in Quebec eventually found their way to Detroit. The Highland Clearances had their parallel in many other parts of what were then the British Isles – the privatisation of common land, the termination of meagre tenancies by greedy landlords and, in Ireland's case, famine, drove millions of 'Brits' off the land and into emigration or the factories. When we speak of 'white privilege' we should never forget this, for the history of many white folk has been far from a privileged one.

As my wife and I turned back from Scarp to get our car from the beach at Hushinish a solitary gannet circled and dived in the bay, and I was reminded of watching a flock of these birds diving in the Sound back in 1971. Exactly this same scene was also captured in a wonderful painting by Norman Adams called the 'Adoration of the Birds'. Evidently Adams came to Scarp to paint for many summers starting in the mid 1960s. He died in 2005 and in a stirring obituary to him in The Guardian *Brian Morley wrote:*

> *his view of nature was perhaps closest to that of Ruskin, who wrote in* Modern Painters *that 'although there was no definite religious sentiment mingled with it, there was a continual perception of sanctity in the whole of nature . . . an instinctive awe, mixed with delight; an indefinable thrill.' In Norman's work, there is always this thrill, a sense of exultation, even in death, and a celebration of both bursting life and inevitable decay, a view of creation which is 'part-pagan, part Christian'.*[5]

PARADISE LOST? The Climate Crisis and the Human Condition

The Adoration of the Birds – Courtesy of the estate of Norman Adams

---

[1] The Scotsman (2017) When the Last Family Abandoned the Tiny Island of Scarp. *The Scotsman,* 12 December 2017. Available at: https://www.scotsman.com/whats-on/arts-and-entertainment/when-the-last-family-abandoned-the-tiny-island-of-scarp-595985 (accessed 21 September 2023).
[2] A wonderful oral history of the Island can be found in the reflections of one of the former residents Donald John Macleod. Macleod, D.J. (2009) *Memories of the Island of Scarp.* Kershader, Isle of Lewis: The Island Book Trust.
[3] The Highland Clearances on the Isles of Lewis and Harris: Some writings of Donald J Macloed of Uig presently of Bridge of Don, Aberdeen. These handwritten writings were sent to Alastair McIntosh and published on his website. See: www.AlastairMcIntosh.com (accessed 1 October 2023).
[4] See: https://sites.rootsweb.com/~qcetcem/gisla_cemetery.htm (accessed 10 September 2023).
[5] Morley, B. (2005) Norman Adams: Painter Driven by a Spiritual Intensity. *The Guardian*, 15 March 2005. Available at: https://www.theguardian.com/news/2005/mar/15/guardianobituaries.artsobituaries (accessed 21 September 2023).

CHAPTER SIXTEEN

STAYING WITH THE TROUBLE

When you come out of disavowal it's usual to get swallowed up by anxiety, grief, guilt or anger, and if this can't be contained, then you may drop into despair. Alternatively, we may find ways of containing these troubling feelings that have suddenly become manifest and, because they are manageable, they can provide the motivational basis for transformation and change. But even when these feelings have been contained they continue to trouble us. We have to learn to face these difficult truths and then stay with the trouble. There's no cure for being human in these times. It's like a chronic condition, it's not going to get better and it may get worse; we'll have to learn to live with it, we'll have to learn how to flourish in spite of it. How *will we* adapt to living in a society where spring has begun to fall silent, where climate refugees besiege the remaining temperate regions of the earth and where austerity is no longer a matter of lifestyle choice but something forced upon us by an increasingly fragile Earth?[1] In other words, how will we adapt to the kind of living that is likely in the Anthropocene if we continue on our 'business as usual' trajectory?

Loss and disillusionment

In my work as a therapist, I almost always find evidence of psychological disturbances in a patient's life which long predate their decision to seek help. These disturbances, signs that the patient's internal reality is out of kilter with their external reality, at first operate beneath the surface manifesting themselves as strange ticks on the face of reality, for example, 'inexplicable' sleeping or eating problems, 'unaccountable' misfortune in personal relationships, or 'unprovoked' rejection from colleagues. Eventually, at a certain point, the individual becomes aware that something is wrong but years may still pass before they accept the need to get some help. But this does not mean that they recognize that they need to change. For change involves personal responsibility and effort whereas cure does not; instead cure offers the promise of magical salvation.

At a cultural level we seek 'cures' for climate change. Sally Weintrobe refers to these as 'quick fix' remedies.[2] They reveal a mindset which paradoxically is both impotent and omnipotent in its hankering after magic. Clive Hamilton has written about these scientific cures when he surveys some of the geo-engineering solutions, which are now part of what are called

'Negative Emissions Technologies'.[3] More recently Elizabeth Kolbert,[4] in her book appropriately named *Under a White Sky,* explores one such quick fix which entails spraying fine particles (like you get in an aerosol) in the stratosphere in an effort to reflect the sun's radiation back into outer space. Of course the problem with these cures – cures which flow directly from our Godlike playing with Earth's atmosphere – is their unintended consequences. As Kolbert notes at the end of her book, each generation of Moderns gets caught up trying to solve problems created by previous generations.

It is often only after different cures have failed to save them that people begin to accept the need to change. So this is the end of the story? No, of course not, because as therapists know from their own personal experience, whilst people may seek change they also seek to stay the same. Change involves loss – of meanings, identities and established ways of relating to self and others – and unless the old can be mourned and relinquished the new cannot be born.

To take the next step requires some real disillusionment. The idea of progress provides a deep sense of meaning for peoples' lives; if this is undermined life can seem meaningless, pointless and without value – reinforcing the nihilism which had already been unleashed by Modernity. If our sense of exceptionalism and entitlement is punctured then this can unleash a powerful sense of grievance, of feeling wronged – how quickly privileged people can feel like victims! So, to facilitate this giving up of illusions, powerful feelings must be contained.

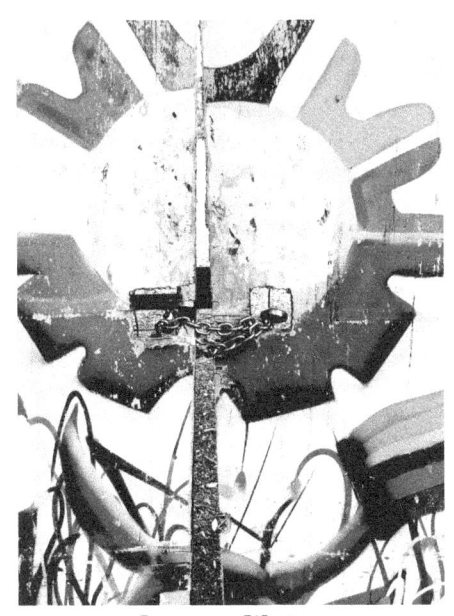

Lawrence Weston

Disillusionment usually doesn't occur through a single moment of insight. Rather, it is the result of a process of struggle which is probably never complete. It would be nice to think that the shedding of illusions is equivalent to an epiphany, like in a fairy tale where the spell is broken and we suddenly wake up. But for the individual in therapy, insight can equally feel like a catastrophe, the very prospect of which can lead to panic and the eruption of more and worse physical and emotional symptoms. Unless this sense of catastrophe can be contained by the therapist, the insights that would break the spell can instead lead to a panicked flight back to the known and familiar, to reaction rather than to emancipation. And so it is with society.

Containing climate distress

When reality bites and people come out of denial, even if they avoid descending into despair, they will be faced with very troubling feelings – fear, guilt, grief, etc – that could be paralysing or could tempt them back into escapism, disconnection or a search for magical solutions. To stay with the troubling feelings we need to be able to contain them. Containment is a key psychoanalytic concept associated particularly with the work of W.R. Bion.[5] It refers to the capacity to stay with and work through or digest emotional experience which is potentially too much to bear and therefore potentially traumatic.

Within the psychoanalytic world much has been written about containment. To contain something psychologically suggests a place where troubling experiences can be taken and shared, the obvious protype being the space created by the arms of the parent who picks up and holds a distressed infant. If all goes well there is a warmth and a steadiness or firmness to this space, this is what makes it safe. If the parent responds appropriately the infant may well have the sense that it has a place in the parent's mind as well as in their arms.[6] Even at this very early, preverbal stage of life I think this is equivalent to the feeling of being understood. From this prototype of the parent's embrace, myriad kinds of containing spaces may be generated – the family home, the school or youth club, a confiding friend, a therapist's consulting room, a good book or film, indeed a society's culture and government. What they all have in common is this: they may offer a place where difficult feelings can be expressed and shared, mulled over and thought about.

Crucially, containment is about finding a space where thinking about these troubling feelings can take place. This doesn't necessarily require another person, the space might be found in nature. Nature can speak to us, it is not just a bundle of our own projections. Here's John Clare in conversation with nature,

> Brazen magpies, fond of clack,
> Full of insolence and pride,
> Chattering on the donkey's back
> Percht, and pull'd his shaggy hide[7]

The mournful sound of a curlew, the awesome power of a wave, the busy persistence of an ant or spider, the crack of a big branch as it snaps from a tree deep in a wood...all such events may cause us to stop, wonder and think.

In relation to the troubling feelings triggered by climate change, one of the findings of my colleagues in the Climate Psychology Alliance is that these feelings are often made worse by the perception that one is alone in having them. Recent research led by Caroline Hickman suggests that for many children it is even worse, that they can feel they are being a nuisance or simply unreasonable by expressing such feelings at home.[8] But something similar can affect adults, including climate scientists. In our interviews with some of the UK's leading climate scientists it became clear that the culture of academic science was itself antithetical to the containment of difficult feelings.[9] It was almost as if scientists were not meant to have feelings about what it was they were studying and so they were left to manage their depression and anger alone.

So knowing you are not alone in having these feelings is an important first step; by normalising them climate change ceases to be the elephant in the room. The environmental and climate movement has pioneered several innovative methods here including Carbon Conversations,[10] Climate Cafes[11] and The Work That Reconnects.[12] All of these involve facilitated groups, sometimes very informal (literally held in cafes), in which people can express and share their feelings about climate change and ecological destruction.[13] They are warm and welcoming, but also secure from intrusion and facilitators are increasingly attentive to the way in which dynamics of class, race and gender can make such spaces feel unsafe for some. Some of these methodologies are now being imported into activist movements such as Extinction Rebellion;[14] some are being adapted for organisational environments such as schools and health services.

So people can find containment for their troubling climate-related feelings in many ways. In my experience, as containment proceeds, people tend to use phrases such as 'I'm more able to navigate my way through' and 'I feel more at peace with my feelings'. Of course, nature itself can provide this containment and some people I know have found that an appreciation of 'deep time' has helped them get a sense of perspective. Others have developed deliberate strategies for switching their attention to the climate emergency on and off so that they 'dose' themselves in manageable proportions. Perhaps all of us need to learn how to disconnect in order to stay connected.

Let me be absolutely clear, containment does not leave us untroubled, as if the wound is healed over, but rather with a 'traumatised sensibility'.[15] Once the reality of climate change and species extinction breaks through it leaves us open and vulnerable, for these realities are not one-off but ongoing, and they are not reparable. As Kassouf puts it, through containment an ability to think and translate thinking into action in a state of precarity is maintained. The world is not assumed to be safe. Containment enables us to hold more, to increase our capacity to face the worst, and be troubled all over again.

Containment facilitates the reconnection of feeling with thinking, providing the motivational basis for engagement with the issue. This is crucial, for the problem with disavowal is that by maintaining a purely distanced and intellectual 'knowing' about climate change, the individual has no motivation to act. Conversely the person overwhelmed by raw, undigested feeling experiences too much disorganisation and paralysis to act. Containment enables us to bring thinking, feeling and action together. I remember one climate scientist saying to us 'action is the antidote to despair' and this is certainly true; agency in itself can have a containing effect related to the sense that one is at least doing something no matter how small. However, it is also true that action can itself become a defence[16] against feeling (and thinking), and in my experience this usually means that such action becomes unsustainable because it leads to burnout.

A final point before leaving this topic - political movements and leaders can themselves provide containment. I say 'can' because of course they (and here I'm specifically thinking about populism) can do precisely the opposite, that is, they can exploit feelings in a parasitical way, particularly feelings like fear, shame and resentment. So what does containing leadership look like? This is a complex issue. Containment can itself become performative, as with the politician who says to his electorate 'I know your pain'.[17] But there are political leaders who have been prepared to confront supporters with difficult truths, to make them think rather than exploit their darker or shallower feelings. One thinks of Vaclav Havel,[18] Nelson Mandela and, today, perhaps Gustavo Petro, Gabriel Boric and Jacinda Ardern.

---

[1] The first widespread intimations of ecological austerity occurred concurrently with the war on Ukraine. The global rise in food prices in 2022, approximating to 10%, were

PARADISE LOST? The Climate Crisis and the Human Condition

only partly attributable to the war. During 2022 the concept of 'food security', along with energy security, was suddenly to be found on the lips of nearly every policy maker. Climate change ensures that it will become an integral long term strategy and not just a temporary response to the war. David, S. (2022) Global Food Crisis: Inflation and Climate Change. BBN Times, 20th June, 2022. Available at: https://www.bbn-times.com/global-economy/global-food-crisis-inflation-climate-change (accessed 4 September 2023).

[2] Weintrobe, S. (2021) *Psychological Roots of the Climate Crisis: Neoliberal Exceptionalism and the Culture of Uncare.* New York & London: Bloomsbury Academic pp.22-6.

[3] Hamilton, C. (2013) *Earth Masters: The Dawn of the Age of Climate Engineering.* New Haven: Yale University Press.

[4] Kolbert, E. (2021) *Under a White Sky: The Nature of the Future.* London: Bodley Head.

[5] Bion, W.R. (1962) *Learning from Experience.* London: Heinemann. I have found two of the most illuminating expositions of Bion's thinking on containment to be found in Meltzer, D. (1978) *The Kleinian Development Part III: The Clinical Significance of the Work of Bion.* Perthshire: Clunie Press. Chapters 4-6; and Britton, R. (1992) Keeping Things in Mind. In R. Anderson (Ed.) *Clinical Lectures on Klein & Bion.* London: Routledge.

[6] Harris, M. (1978) Towards learning from experience in infancy and childhood, in *The Collected Papers of Martha Harris and Esther Bick.* Strath Tay, Perthshire: The Clunie Press.

[7] *Selected Poems of John Clare*, edited with an introduction by James Reeves. 1954. London: Heinemann, p. 8.

[8] Hickman, C. et.al. (2021) Climate Anxiety in Children and Young People and their Beliefs About Government Responses to Climate Change: A Global Survey. *The Lancet*, December, 2021. This research surveyed 10,000 children across the globe, north and south, and found that children experienced two main sources of stress – the ecological destruction occurring all around them, and the failure of the adult world to take appropriate action. As Caroline put it, many children felt that adults, including their own parents, simply didn't 'get it', i.e. simply didn't get the rawness of the feelings that they had.

[9] Hoggett, P. & Randall, R. (2018) Engaging with Climate Change: Comparing the Cultures of Science and Activism. *Environmental Values*, 27: 223-243

[10] Randall, R. & Brown, A. (2015) *In Time for Tomorrow? The Carbon Conversations Handbook.* Stirling: The Surefoot Effect.

[11] Broad, G. (2024) 'Ways of being' when facing difficult truths: Exploring the contribution of climate cafes to climate crisis awareness. In Anderson, J., Staunton, T., O'Gorman, J. and Hickman, C. *Being a Therapist in a Time of Climate Breakdown.* London: Routledge.

[12] The Work That Reconnects (2022). Available at https://workthatreconnects.org (accessed 4 September 2022).

[13] A valuable summary of a variety of these approaches has been provided by CPA colleagues. Tait, A., O'Gorman, J., Nestor, R. & Anderson, J. (2022) Understanding and responding to the climate and ecological emergency: The role of the therapist. *British Journal of Psychotherapy*, 38(4): 770-779.

[14] This is embodied in the concept of the 'regenerative culture', one of XR's stated core values.

[15] Kassouf, S. (2022) Thinking Catastrophic Thoughts: A Traumatised Sensibility on a Hotter Planet. *American Journal of Psychoanalysis*, 82: 60-79.
[16] Psychotherapists refer to this as 'enactment' or 'acting out'.
[17] Many modern, liberal, technocratic political leaders such as Clinton, Blair and now Macron have been able to draw artfully upon aspects of therapeutic culture. See my examination of what I call 'the modern charismatics'. Hoggett, P. (2009) Politics, Identity & Emotion. Boulder, Col: Paradigm. Chapter 3.
[18] Havel, the banned playwright come new president of Czechoslovakia reflected on being in power in his Summer Meditations. He offered these critical reflections whilst still in office, the first essay 'Politics, Morality, and Civility' is particularly thought provoking. Havel, V. (1992) *Summer Meditations*. London: Faber and Faber.

CHAPTER SEVENTEEN

THE TRAGIC POSITION

Following Bill McKibben, I argued in Chapter 15 that in the face of climate breakdown, we are forced to recognise that we have embarked upon a long descent, the final potential destination being hell, but there are many stops on the way, and we must fight for all we are worth for the least worse outcome. The tragedy of climate change lies not just in each life lost, nor even in the loss of whole natural systems of life, but in the wanton nature of this loss and in its irreversibility.

Human fragility

In their last years my parents had moved to live close to us in Bristol. Their new flat had a wonderful view as it looked north up the Severn Estuary at Portishead, a small town on the Bristol Channel. The closeness to the sea reminded my father of his early life, his father having been a trawlerman fishing out of Lowestoft.

In the last months before his death my father shared a dream with me. He was back at the factory where he'd got his first engineering job and was due to appear before the 'big chief'. He'd failed his exam and the implication seemed to be that he might lose his job. But the summer holidays intervened and after they were over, he went back to work only to find that his desk had been cleared and there was someone else working there. What was worse was that none of his colleagues, not even his closest friend, recognised him. He had become invisible to them.

It is with regret now that I realise I lacked the courage to speak to the sadness and fear in my father's dream. I remember us laughing together at how bizarre dreams could be in an unspoken denial of the meaning it was inviting us to share and explore. Time moves differently when you are waiting to go to see the big chief. We're not in any hurry until the pain becomes intolerable. Many friends and acquaintances have told me of loved ones who literally seemed to 'slip away'. My hunch is that we are, mostly, called to go when our time has come, and we can either respond quietly to that calling or resist it.[1] Modern medicine seems designed to enable us to resist it.[2]

And so who is this big chief? God? The grim reaper? Surely it is none other than our own nature, that creature inhabiting us, calling us. We don't have to be near death to experience this. In ordinary sickness we feel oppressed by our bodies. As I mentioned in Chapter 2, for the Italian philosopher Sebatstian Timpanaro this brings to our attention what he calls the 'passive aspect' of our relation to nature.[3] We may spend much of our time oppressing nature, manipulating her to serve our purposes, an active domination, but ultimately it seems she triumphs over us. The question she puts to each of us is this: can we surrender to her with grace and honour?

Photo courtesy of Owen Newman

For some, the answer is a negative. Do not go gentle into that good night...rage, rage against the dying of the light.[4] It occurs to me that this rage could have two very different provocations. Where life gives us much joy the prospect of it ending fills us with sadness and inclines us to protest, not at the fact of death but at its timing. Then there is another possibility, that the rage is provoked by a sense of lack rather than fullness, of things not yet done, of a life somehow incomplete and unfinished, and I think this is where the resentment comes from, as if to die is to be cheated. Like many others who have had a life-threatening illness, when I got a cancer diagnosis I was surprised by my equanimity. What was unbearable was the period of 'not knowing' before the diagnosis. Nevertheless, the thought of that last journey is still likely to arouse demons in our unconscious. One day after a death of someone dear to me I dreamt of a huge coiling sea monster in the harbour at Lowestoft. There were crowds upon the quay and for some unaccountable reason a boy jumped into the water. We knew in horror what that meant. I woke up.

This dying of the light is the shadow that (our) creaturely nature casts upon us, a shadow against which we are ultimately defenceless. It signals the immanence of human frailty and vulnerability. Freud referred to it as the hand of Fate. So we do not need an exotic cosmology to understand that we are of nature, and that nature is not set apart from us. We are nature, the fingers that type these words, the eyes that look at them, they are of nature, human nature.

The capacity for care

So when we speak of nature, of our alienation from nature, of our desire to master nature, of nature as this foe which must be conquered, are we necessarily speaking of mother nature, that nature 'out there', the nature of trees and seas? Or are we also speaking of the nature 'in here', the nature that resides within us? Does not this second nature seem like an unnoticed intruder, an unwelcome guest who might, without a moment's notice, upset the peace? Are we not equally alienated from this nature, set upon a desire to conquer it and deny our actual enthrallment to it?

Surely it follows that there must be an intimate connection between the two struggles, the struggle to reintegrate the splits between the modern self and external nature, and the modern self and our vulnerable bodily nature, that is, our struggle to overcome this separation from nature both without and within. Perhaps they go together. Perhaps there can be no peace with external nature without our also being at peace with our internal nature, that is, our nature as natural, creaturely beings.

Upon this terrain of the body, this place of frailty, an ethics has grown.[5] It has emerged from the patient being-with of the one who cares for the vulnerable other – at first the helpless child, much later the equally helpless elder. What has become known as the 'feminine ethic of care' places huge emphasis upon the relational dimension of care, upon the quality of relationships rather than the quantity of inputs.

I never met Peta Bowden but was much moved by some of the themes in her book *Gender Sensitive Ethics*[6] which greatly influenced an essay I wrote called 'Hatred of Dependency' over twenty years ago.[7] At that time I was barely conscious of climate change. All my writing and research focused on the nature of welfare and our ideas of a welfare society and the welfare state. I was very exercised by the absence of any reference to the emotions in theory and research in social policy, and in the social sciences more generally. This was before the so-called 'emotional turn' in the social sciences. Peta was one of the few people around who seemed to know intuitively how important the emotions were in relations of care in nursing and other welfare practices.

Two decades later and my close colleague Sally Weintrobe began developing the idea of 'frameworks of care' in relation to climate change.[8] This prompted me to revisit the writing on care which had once been so influential for me. In gathering my thoughts I was curious to find out more about Peta. Was she still working and writing about the nature of care? I was aware that I didn't even know where she lived and worked. I thought I might quite like to look her up on the internet. So it was with sadness that I discovered that she had died in 2018. I discovered this via tributes from her colleagues at Murdoch University, Perth, in Australia, which were posted on the web. There is one lovely tribute that provides me with an opportunity to introduce Peta, to bring her back to life as it were:

> Dr Peta Bowden, outstanding human being, dear friend, academic, teacher, activist and former member of Murdoch's Philosophy Program, passed away on Wednesday 6 June 2018.... Peta's care for others came from her deeply-felt knowledge that we are part of a sentient 'kind'. This knowledge was expressed in her compassionate outreach to all she encountered, whether as part of the human or 'other-than-human' world. A close friend and fellow academic remembers Peta watching, entranced, as a small group of baby ducks crossed a roadside puddle, and recalls her close attention to the devastations that we, as humans, cause to our planet. With this intimation coming from her sense of the shared human condition, Peta was always, and most singularly, a *compassionate* campaigner. This orientation infused all her actions, whether campaigning for the Greens, undertaking detailed work for 350 Australia, or advocating for the value of the humanities at Murdoch.

I was so delighted to hear that Peta's passion had extended to the other-than-human and had informed her environmental activism. And the phrase, that we are part of a sentient 'kind', got me thinking some more. The dictionary defines 'sentience' as the capacity to feel, perceive or experience subjectively, and of course for Peta the sentient 'kind' that we are part of stretches out way beyond the human species to include baby ducks, probably trees and perhaps Gaia itself. After all, the Earth is like a living planet, very probably not unique to the cosmos, but probably unique to this corner of it. There is a solidarity figured here – all that feels also suffers.

During the coronavirus pandemic I noticed that I received a number of messages which ended with the phrase 'take care'. It occurs to me that there are so many meanings wrapped up in this phrase. The first and obvious one is that it is an expression of concern and affection from the sender, that they 'care about' us. The second meaning that comes to my mind is to be 'careful', that is vigilant and watchful in a non-paranoid way. A third meaning is to be 'care-full', that is to keep one's capacity for care topped up and full (for care is essentially a capacity to be developed and not a skill to be learnt). There is also a useful distinction between 'caring about' and 'caring for'. To 'care about' someone is to keep them in mind (care as thought-fullness, something central to what therapists do), whereas to 'care for' someone also involves the more direct sense of looking after someone who is sick or vulnerable. As Peta Bowden understood, care is an ethical practice involving the moral integrity of both carer and cared for.[9]

Note that with the exception of 'caring for', virtually none of these dimensions of care are quantifiable. Even the work of 'caring for' involves both relational and emotional engagements with the other that are elusive and crucial precisely because of this. But for the performative cultures that now govern both corporate and governmental sectors in neoliberal societies what counts is what can be counted. Most of what comprises care simply doesn't count and nor do the care providers. As the COVID-19 pandemic temporarily challenged a number of fundamental neoliberal beliefs, we suddenly discovered that this army of unnoticed and underpaid workers were 'essential workers'.

The struggle to create a welfare society built upon an ethic of care and the struggle to prevent environmental destruction and climate chaos are not separate struggles. They share much in common and need to draw upon each other's sensibilities and practices much more than they do at the moment.

Tragic exceptions

Joel Kovel, who worked as a psychoanalyst in the 1970s and ran for the USA Green Party's Presidential nomination in 2000, once said that we must resist those who would sever humanity from nature as much as those who would submerge it in nature.[10] I sometimes feel as if I am being asked to choose: are you pro nature or pro human? Looming behind this 'choice' one sometimes finds two abstract concepts, ecocentrism and anthropocentrism. There are many ways of distinguishing between the two. One distinction that I have no problem with refers to whether we see nature as having an intrinsic value or whether we see it essentially as something put here for human purposes (instrumental value). If one subscribed to the latter one could still be 'green' but this would be an anthropocentric form of green ethics – the religious notion of 'stewardship' is a good example. If this was the only difference between the two orientations, then I would put myself firmly in the ecocentric camp. As we saw in Chapter 1, according to the Book of Genesis, God gave man dominion of nature – we should have no truck with this kind of exceptionalism.

Nor do I have a problem with a notion of subjectivity that includes both the human and non-human. Stephen Duguid[11] illustrates this perspective with a beautiful extract from Aldo Leopold's *A Sand County Almanac:*

> It is in midwinter that I sometimes glean from my pines something more important than woodlot politics, and the news of the wind and the weather. This is especially likely to happen on some gloomy evening when the snow has buried all irrelevant detail, and the hush of elemental sadness lies heavy upon every living thing. Nevertheless, my pines, each with his burden of snow, are standing ramrod straight, rank upon rank, and in the dusk beyond I sense the presence of hundreds more. At such times I feel a curious transfusion of courage.

Here then is nature as subject, speaking to us if only we had the ears to listen. And with this status of subject surely the other-than-human must also be a rights-bearing subject with all the implications this entails for domesticated and wild animals, indeed for whole ecosystems such as coral reefs, rivers, etc.

But on the side of anthropocentrism I do believe that there is an ontological distance (but not divide) between the human and the non-human. Whilst we are all subjects, the human is a particular kind of subject. Whilst I am convinced by the evidence that higher animals including apes, elephants, whales and dolphins have language and culture,[12] have the capacity to identify with others, and have feelings such as sorrow, grief and compassion, it would be a giant leap from here to say that they also are aware of their own mortality, or have symbolic universes such as science and the arts through which they can develop understanding of the nature of their own existence (self-understanding) and of the world in which they live. Nor, as far as we know, do they engage in religious practices, prostrate themselves before Gods, worry about the passing of time, imagine they have superhuman powers, fret about leaving a mark upon the earth after they die or any of the other practices that we modern humans engage in because of our mortality consciousness.[13] My argument is that, far from being necessarily hubristic and triumphalist, one can also imagine an anthropocentric perspective which was deeply tragic, one where we moderns are still seen as an animal but a rather tragic and exceptional animal because of the suffering that we cause to ourselves and to all others, human and nonhuman.

We are a contradiction. Subjectivity, manifest in its highest human form, is a strange outgrowth of nature through which one part of nature has developed the capacity to become self-aware, take itself as an object of contemplation and shape itself in a conscious way. And yet it is still of nature – human subjectivity remains trapped within the confines of the body. Perhaps one day will come when it will break free, when the replicants will take our place and/or the transhumanist project is completed. But until that day we have to live this contradiction.

This means that our subjectivity, a unique form of subjectivity, is both the source of our liberation and a curse. It is our curse because to be human means to have to bear loss. Every second passing, every sound fading, every smile received and embrace offered, every plan put into action, every sight and sound and smell and touch of our lives is transient. And we know this too and so we also know of our own passing and the passing of those that we love. We bear these losses every day and on good days because of this we are uplifted by a love of life whilst on bad days we feel despair and terror.

Transience

Earth and spirit: it is from these antinomies that humankind is formed. Walking together in the summer of 1913 Sigmund Freud and the young poet Rainer Maria Rilke were engaged in a conversation which became the subject of Freud's famous paper 'On Transience'.[14] Freud noted that Rilke admired the beauty all around them but felt no joy in it because 'he was disturbed by the thought that all this beauty was fated to extinction'.[15] Faced with the transience of all sentient life, Freud felt that Rilke was overcome by an 'aching despondency'.[16] Freud also noted the existence of an alternative response, a defiant refusal to believe that the world in which we are all immersed will fade away to nothing. But this demand for immortality was a refusal to face the pain of loss, the pain that had temporarily overcome Rilke. Freud adds that he disputed the poet's view that the transience of all things destroyed their value; to the contrary, it was precisely their transience that gave them their value.

It is curious to observe how, in the years after their famous conversation on the subject of transience in 1913, Freud and Rilke almost appeared to change positions. After his experience of the First World War, Freud became increasingly pessimistic and Rilke less despondent. Joanna Macy, the great philosopher and environmentalist, has written a wonderful introduction to a selection of some of Rilke's later work, the *Duino Elegies* and *Sonnets to Orpheus,* in which she traces Rilke's psychological and creative recovery from the horrors of the war.[17] One hundred years on and once more on the brink of destroying our world, Macy suggests that Rilke shows us the direction we need to go by 'offering ways to dignify our pain for the world and deepen our capacity for gratitude'.[18] Like Rilke, we must look at the full horror of what we are capable of, be prepared to be dismembered (torn apart) by it, in order to find the grounds for a renewed creativity. Like Orpheus we must look into a darkness which is gathering both out there and in us in order to nurture our love for the world.

Macy argues that Rilke 'affirms the transformative power of loving what must disappear'.[19] She insists that this sense of impermanence deepens our identification with the natural world whose presence, like ours, is so ephemeral. Mortality, rather than being some unknowable end point, inhabits our very being, our fading from view is simultaneous with our coming to be.

Rilke's poetry is rich in organic metaphor, pitching the human spirit into the earth, the great web of life, from which it has sprung and to which it will return. Macy ends her fine introduction thus:

> What gives the Duino Elegies and the Sonnets to Orpheus their visceral potency is Rilke's determination to cherish life in direct confrontation with what robs us of life. In the face of impermanence and death, it takes courage to love the things of this world and to believe that praising them is the noblest calling. Rilke's is not a conditional courage, dependent on an afterlife. Nor is it a stoic courage, keeping a stiff upper lip when shattered by loss. It is courage born of the ever-unexpected discovery that acceptance of mortality yields an expansion of being. In naming what is doomed to disappear, naming the way it keeps streaming through our hands, we can hear the song that streaming makes. Our view of reality shifts from noun to verb. We become part of the dance.[20]

Transience, fragility, loss, beauty, care...these are the fundaments of that love for the world whose traces can be found both in ecological and welfare practices.

The tragic position

Nearly forty years ago the psychoanalyst Neville Symington asked us to consider the existence of a state of mind he called the 'tragic position', one which confronts the abyss of non-meaning.[21] I believe that if we are to take up the role required of us in our climate emergency then we need to more fully embrace this tragic position.[22] Leon Wurmser very usefully suggests a fourfold layering of the dynamics of tragedy.[23] The first concerns the frightening

PARADISE LOST? The Climate Crisis and the Human Condition 145

intensity of the tragic compulsion to act self-destructively that sometimes grips an individual or society. The second and third concern the force of irreconcilable commitments, loyalties and values which can tear the self or the group apart particularly where such values are invested with an absolutism so that they are held to with an implacable stubbornness or sense of narcissistic correctness. Tragedy arises not when evil triumphs over good but when two goods clash and destroy each other, not when destructive acts occur which take generations to repair but when destruction occurs which is irreparable. Tragedy concerns that which is irrevocable, and irredeemable, that which has no resolution and therefore no possibility of healing.

Finally, the fourth level, Wurmser believes tragedy lies at the very heart of the human condition. We are a contradiction, the one species on the earth which is both of nature and yet beyond nature. Wurmser likens it to a massive trauma, an ancient woundedness.[24] As humankind lifted itself above the struggle for bare survival it began to become increasingly aware of this tragic contradiction that inhered to being human. We Moderns know of our transience and vulnerability and yet we cannot quite believe what we know. Our Promethean drive to master the universe is a manic defence against this knowledge and the terror that it elicits in us. We *will* become Gods. Every new scientific discovery, every extension of our control over the human and other-than-human, in short, every manifestation of our hubris is a flight from this unthought and unthinkable known. We, individually and collectively, are born of nature and will soon return to nature, that our era (the era of the human) is but a gnat's blink in the life of this universe, and that nature with all its limits and constraints is closing upon us, always closing upon us. This fact of life seems so hard for us Moderns to embrace.

---

[1] A theme which is very present in the fictional encounter between the anthropologist W.H. Rivers and the village elder and medicine man Njiru in *The Ghost Road*, the third book in Pat Barker's *Regeneration* trilogy. Barker, P. (1995) *The Ghost Road*. London: Viking. Bob Simpson's review of the anthropological evidence of the way in which death has been handled in different cultures provides ample evidence of the contrast between premodern and modern societies. Simpson, B. (2018) Death. In *The Open Encyclopedia of Anthropology*, ed. Felix Stein. Munich: Open Knowledge Press.

[2] Bill McKibben devotes several chapters in his recent book *Falter* to attempts by Silicon Valley biotech firms to do precisely this.

[3] Timpanaro, S. (1970) *On Materialism*, trans Lawrence Garner. London: New Left Books.

[4] Thomas, D. (1952) Do not go gentle into that good night. *Collected Poems 1934-1952*. London: Dent.

[5] Bowden, P. (2000) An 'ethic of care' in clinical settings: Encompassing 'feminine' and 'feminist' perspectives. *Nursing Philosophy*, 1(1): 36-49.
[6] Bowden, P. (1997) *Caring: Gender-Sensitive Ethics*. London: Routledge.
[7] Hoggett, P. (2000) Hatred of Dependency. In *Emotional Life and the Politics of Welfare*. London: Macmillan.
[8]. Weintrobe, S. (2021) *Psychological Roots of the Climate Crisis: Neoliberal Exceptionalism and the Culture of Uncare*. New York & London: Bloomsbury Academic.
[9] See Bowden, Caring, note 6 above.
[10] Kovel, J. (2002) *The Enemy of Nature: The End of Capitalism or the End of the World*. London: Zed Books.
[11] Duguid, S. (2010) *Nature in Modernity: Servant, Citizen, Queen or Comrade*. New York: Peter Lang Press pp. 10-11.
[12] Ball, P. (2022) *The Book of Minds: How to Understand Ourselves and Other Beings, From Animals to Aliens*. London: Picador.
[13] Of course, what I am talking about here is the neurotic dimension of modern civilization. Now whilst it is possible that some animals may exhibit neurotic behaviour like some of Pavlov's experimental animals their neurosis is undoubtedly a result of their contact with humans. Norman O. Brown makes this point when arguing that there appears to be an essential connection between being sick (psychologically) and being civilized (which of course was also Freud's position). Brown, N. (1959) *Life Against Death: The Psychoanalytic Meaning of History,* Routledge and Kegan Paul. Chapter 7.
[14] Freud, S. (1915) On Transience. Standard Edition of the Complete Psychological Works of Sigmund Freud, Vol. 14, pp.303-308. London: Hogarth Press.
[15] Ibid, p. 303
[16] Ibid. p. 305
[17] Barrows, A. and Macy, J. (2016) *Rainer Maria Rilke: Selections from Duino Elegies & Sonnets to Orpheus*, trans Barrows & Macy. Battleboro, Vermont: Echo Point Books & Media.
[18] Ibid, p. 6.
[19] Ibid, p. 11.
[20] Ibid, p. 23.
[21] Symington, N. (1986) *The Analytic Experience*. London: Free Association Books. pp.274-276.
[22] The philosopher John Foster makes this case powerfully in Foster, J. (2022) *Realism and the Climate Crisis*. Bristol: Bristol University Press.
[23] Wurmser, L. (2015) Mortal wound, shame, and tragic search: Reflections on tragic experience and tragic conflicts in history, literature and psychotherapy. *Psychoanalytic Inquiry*, 35 (1): 13-39
[24] Ibid, p. 18.

SECTION FOUR

LESS IS MORE, A NEW ETHIC

Hartland looking south

It is early June and cumulus clouds are swelling above the landmass of Devon and Cornwall. But the Atlantic is free from cloud and this afternoon the sea dazzles for what seems like hundreds of miles out to the west. The remote and wild coastline which stretches south from Hartland Point is my favourite place in the world, a rolling succession of high cliffs and deep valleys which continue for a dozen or so miles until you meet Bude's long strand. I am on a cliff about 450 feet above a deep valley which flows into the Atlantic at a spot called Duckpool. Far to the south I can see a series of headlands – Tintagel, Pentire Point, Trevose Head, and perhaps even the Land's End Peninsular. A few miles south from where I am stood, the retreating tide has revealed Bude's beach touched by lines of breakers; I can just make out clusters of surfers and beach walkers.

There's a smile enveloping my entire body as I walk the cliff path, and I get to thinking. If I knew that I would never see such beauty or experience joy like this again, would I go quietly into the night or would I rage against the dying of the light? A sadness falls upon the landscape at this thought and I realise that far from timeless, what I see before me now is very different to what I would have seen just one hundred years ago. Everything passes, or so they say, but what about that dazzling sea or the wind in my ears whistling up from eons past? This latter thought cheers me up and reminds me of how rich, how wonderful, sadness is.

I walk on a bit and suddenly I find myself in the future and full of foreboding. My head tells me that the coming decades will not be easy for staying human. So what must we put in our ruck sacks if we are to ensure that that the coming troubles bring out the best rather than the worst in us?

CHAPTER EIGHTEEN

PARADOXES OF HOPE

False hope

In Chapter 15 I noted that, regarding the climate emergency, the certainty of despair is often preferred to the precarity of hope. I have also noticed how, with some of the suicidal people I work with as a therapist, there is an insistent belief that their difficulties cannot be overcome and that there is nothing to be done. Reflecting on her work with traumatised refugees the psychoanalyst Alessandra Lemma noted that this kind of certainty in one's mind that nothing will change is the antithesis of what she calls 'mature hope'.[1]

When some of my patients slip into despair their sense of their strengths, that is, of their internal resources, evaporates. This is sometimes manifest literally and bodily - they become exhausted, even minor exertions leave them drained. What were small obstacles become insurmountable problems, molehills became mountains. Their sense of their own agency evaporates. It is as if their world lies devastated and in ruins and they can do nothing about it.

Lawrence Weston

Despair can be immobilising and terrifying. It arouses powerful defences. Paradoxically one of these is a particular kind of hopefulness, one where hope is clung to in desperation. Anxiety and fear underly this kind of false hope. In everyday language we talk about 'pinning our hopes' upon something, upon some outcome that is desired. This kind of desperate or fearful hope is also manifest in political situations, like with the growing climate emergency. False hope ties itself to an outcome. All our efforts become geared towards this outcome, we do all that we can to make it happen, thinking that if we just push a little harder the desired effect will be achieved, as if political action was a simple cause/effect relationship. As the American writer and activist Rebecca Solnit noted, it is for this reason that some activists 'specialize in disappointment'.[2]

Here is an example. I sometimes run workshops called 'Taking the Heat out of Talking About Climate Change' which focus on how to go about talking to friends, family and colleagues about climate change and our collective failure to take requisite action. One of the key principles is not to enter such conversations with an outcome in mind, ie. to change the other person in some way, but to enter with a spirit of curiosity in, and respect for, the other. For I know from experience that if I go with an outcome in mind in no time at all the 'conversation' has become an argument. In pinning my hopes upon achieving change in the other, I reveal the extent to which my intent is a defence against my own anxieties and uncertainties.

Facing the worse

Paradoxically perhaps, the worst is what the hopeful are always prepared for. Facing the worst (pessimism of the intellect), and yet sustaining an optimism of the will, now there's a challenge. Facing climate change, species extinction, global conflicts and poverty, allowing ourselves to be disturbed by them, moved and even dismembered by them, and yet remaining sane, is no easy thing. But the hope that comes from being able to face the worst is an enduring hope because it is not built upon a scaffolding of illusion and wishful thinking. It is defiant and courageous and it refuses to capitulate to what might seem like hopeless odds. This kind of hope is something we do rather than have, say Joanna Macy and Chris Johnstone in their book *Active Hope*.[3]

Let's face it, we don't know what the future holds. When people begin to emerge from the ruins of a life they can see painfully what has been lost but ahead is only an uncharted sea. But move on they do, and without false hopes. Some kind of elemental confidence about life is slowly restored. Followers of the psychoanalyst Melanie Klein believe that hope is equivalent to trust in the existence of goodness.[4] This trust arises only if a person is able to face their own destructive attacks on what is good and, through the experience of guilt, mobilise their love to repair what has been damaged. At the cultural level this is akin to recognizing the worst that we are collectively capable of and using that recognition as the inspiration to make amends.

The containment offered by the therapeutic environment provides the self with the strength to face difficult truths and in facing them the self becomes stronger. I have learnt to have enormous respect for people's capacity to face the worst when they feel they have the permission. It might sound odd to say this but more often than not people need to be given this permission, they need to feel that they can visit the despair with someone who does not need to be protected from it. And to really face the worst in this way paradoxically often brings relief. By naming it, what had previously been a source of internal terror begins to lose some of its sting. One thinks of Roosevelt's famous phrase 'there is nothing to fear but fear itself'.

Mature hope

When we set up the Climate Psychology Alliance in 2012 it was no coincidence that our logo read 'Climate Psychology: Facing Difficult Truths'. But in recent years I have come to reflect on this, seeing in the strapline a kind of psychoanalytic stoicism[5] which tends to give too much emphasis to 'bearing reality' and 'facing the facts'. For the reality is that reality is not immutable.

Rebecca Solnit notes that uncertainty is the soil in which real hope flourishes, she calls it 'the embrace of the unknown and the unknowable'.[6] In a similar vein Lemma defines mature hope (as opposed to false hope) as 'a state of mind of expectant possibility that is rooted in reality'.[7] An association forms in my mind; when someone is pregnant we often say 'she's expecting'. Perhaps then, a hopeful state of mind is one pregnant with possibility.

I'm immediately reminded of the paediatrician and psychoanalyst D.W. Winnicott's idea of 'potential space'. Winnicott was concerned with meaning, play and what he called a creative engagement with reality and fact, as opposed to one simply of acceptance and compliance.[8] Repeatedly in his reflections on play he referred to it occurring in 'potential space' – it is the pliable space where meaning is generated, culture is formed and politics unfolds. This state of mind of expectant possibility is perhaps what mature hope really is.

Culture and politics take place in the space of the 'not yet', of the 'what might be'. Those in the business of art, film and literature know that it is not facts that move people but meanings, imagery and stories. Following the gay activist and scholar Jeffrey Weeks[9] I suggest we think of these as 'fictions' and in politics we can think of slogans like 'Think Globally, Act Locally' or 'There is no Planet B' as mobilising fictions[10] designed to change reality by galvanising social and political movements. Rather than a stoical attitude of acceptance this corresponds to a rejection of 'what is', a refusal to accept the givenness of reality.

As we shall see in Chapter 24, in democratic societies these competing fictions, these competing ideas of what is good, are able to collide, engage, contest, fracture and evolve within a framework of principles, practices and institutions which holds and contains them. This has been explored by feminist political philosophers such as Iris Marion Young and Seyla Benhabib.[11] This framework of principles, practices and institutions which sustains democracy is what Sally Weintrobe terms a 'framework of care'.[12] I think Winnicott would call what goes on within this framework as 'shared playing'. As we have seen in Section 2 of this book, it is this crucial political and social achievement (i.e. democracy) which is presently under threat, from both the inside and the outside.

Radical hope

What happens in times like today, times of enormous social discontinuity, where the taken-for-granted ongoingness of the world is suddenly placed in doubt? This is the subject of Jonathan Lear's book *Radical Hope: Ethics in the face of Cultural Devastation.*[13] Lear focuses on what happened to the Crow tribe in North America when white settlers in the late nineteenth century eradicated the buffalo around which the Crow's entire way of living revolved. Lear notes how some Crow, like the medicine woman Pretty Shield, continued with cultural practices like cooking for warriors before battle even though the Crow had been disarmed and corralled into a reservation. Lear likens this to the repetition-compulsion of the melancholic. Others looked for magical or messianic solutions or exhorted fellow Crow to follow them in fighting the white settlers – a path to almost certain extermination.

But the Crow Chief, Plenty Coups, who adopted none of these paths is the object of Lear's study, for Plenty Coups chose neither denial nor despair but what Lear calls 'radical hope', that is, the capacity to imagine some way in which the Crow could go on being Crow even after all that had given meaning to their life had been destroyed. Lear notes how the Crow were able to use their imagination, and particularly their interpretation of dreams, as a means for facing a future which at that point was beyond their comprehension. And Lear draws the parallel with psychotherapy for, he argues, the therapeutic stance involves 'hopefulness toward a future self and a future way of being that cannot yet be fully comprehended'.[14] Lear adds that 'there

PARADISE LOST? The Climate Crisis and the Human Condition

is a trust that the world has something good in it, even if we do not yet have the resources to grasp what it is'.[15]

In his book Lear depicts a whole community, the Crow Indian in late nineteenth century North America, coming to terms with loss of a way of life in this fashion. Lear calls this 'radical hope', a commitment to the idea that the goodness of the world transcends one's limited and vulnerable attempts to understand it. So radical hope is not just about determination and courage, it is also about love and a rediscovery of all that is benign in the world.

Our modern way of living presently stands close to the edge of catastrophe just as the Crow's did. Lear says, 'at a time of cultural devastation, the reality that a courageous person has to face up to is that one has to face up to reality in new ways'.[16] By squarely facing up to the end of civilization as we know it might something new, entirely beyond our current imagination, come into being? Civilization, but not as we presently know it?

The power of hopelessness

Vasily Grossman based his stunning novel *Life and Fate* on his experiences as a Red Army war correspondent.[17] He was present at the battle for Stalingrad, witnessed Nazi ethnic cleansing in Ukraine and was one of the first journalists to enter Treblinka. Like other writers, such as Primo Levi and Hannah Arendt, who survived the devastation of Nazism, Grossman noted how the desperate hope that one's own life, or the life of one's loved ones, would be spared made many of the Nazi's victims obedient and compliant. It worked to disarm people.

Grossman and Arendt noted the paradox that the most courageous struggles, such as the Warsaw Rising,[18] 'were all born of hopelessness'.[19] His fictionalised account of the fighters in 'House 6/1' at Stalingrad, fighters who know they will not survive and yet struggle on with tenacity, humour and integrity, is his testimony to the power of hopelessness. It follows that on occasions hope can be the enemy of will and determination. Perhaps then we invest too much hope in hope, perhaps what enables some individuals and groups to struggle on when facing apparently insurmountable difficulties is to have courage and determination in the absence of hope. Perhaps then there is something to be said for the practice of being hope-less. Perhaps this is what Gramsci meant by 'optimism of the will', the focus for Chapter 20.

---

[1] Lemma, A. (2004) On Hope's Tightrope: Reflections on the Capacity for Hope. In Susan Levy and Alessandra Lema (Eds.) *Perversion of Loss: Psychoanalytic Perspectives on Trauma.* London: Routledge.

[2] Solnit, R. (2005) *Hope in the Dark: Untold Histories, Wild Possibilities*. Edinburgh: Canongate Books. p.60.
[3] Macy, J. and Johnstone, C. (2012) *Active Hope: How to Face the Mess We're In Without Going Crazy*. Novato, Calif: New World Library.
[4] Klein, M. and Riviere, J. (1964) *Love, Hate and Reparation*. New York: W.W. Norton & Co.
[5] Hoggett, P. (2022) Climate Change: From Denialism to Nihilism. In Hollway, W., Hoggett, P., Robertson, C. and Weintrobe, S. *Climate Psychology: A Matter of Life and Death*. Bicester: Phoenix Publishing House. pp.40-1.
[6] Solnit, *Hope in the Dark*, note 2 above, p. xii.
[7] Lemma, *On Hope's Tightrope*, note 1 above, p.125.
[8] Winnicott, D.W. (1971) *Playing and Reality*. London: Routledge.
[9] Weeks, J. (1991) *Against Nature: Essays on History, Sexuality and Identity*. London: Rivers Oram Press.
[10] Hoggett, P. (2000) Mobilising Fictions. In *Emotional Life and the Politics of Welfare*. London: Macmillan.
[11] See for example the collection of essays (including one from Iris Marion Young) in Benhabib, S. (1996) (ed.) *Democracy and Difference: Contesting the Boundaries of the Political*. Princeton, NJ: Princeton University Press.
[12] Weintrobe, S. (2021) *Psychological Roots of the Climate Crisis: Neoliberal Exceptionalism and the Culture of Uncare*. New York and London: Bloomsbury Academic
[13] Lear, J. (2008) *Radical Hope: Ethics in the Face of Cultural Devastation*. Cambridge, Mass: Harvard University Press.
[14] Ibid, p. 301.
[15] Ibid, p. 304.
[16] Ibid.
[17] Grossman, V. (2006) *Life and Fate*, trans Robert Chandler. London: Vintage;
[18] Samantha Rose Hill notes that Hannah Arendt saw the Nazis' ability to exploit fear and hope as meticulous. It facilitated Jewish passivity and compliance in the Warsaw Ghetto. It was only when they gave up hope and let go of fear that they realised armed resistance was the only moral and political way out. Samantha Rose Hill (2021) When hope is a hindrance. AEON, 4th October 2021. Available at: https://aeon.co/essays/for-arendt-hope-in-dark-times-is-no-match-for-action (accessed 6 September 2023).
[19] In a moving passage Grossman (see note 17 above, pp. 197-200) reflects on hope, obedience and the power of hopelessness.

CHAPTER NINETEEN

WHAT IS THIS THING CALLED LOVE?

Learning from Etty

Etty Hillesum was a young Jewish woman from Amsterdam who worked in the hospital at the Westerbork Transit Camp during the Nazi occupation. She travelled between Westerbork and Amsterdam several times on behalf of the Jewish Council, refusing offers from Dutch friends to help her escape. Although she didn't know about the gas chambers, she'd intuited that Dutch Jews probably faced extermination and she saw it as her calling to be with her fellow Jews in their suffering. After eight months at Westerbork her own turn came and she was shipped off to Auschwitz in September 1943, never to return.

A psychoanalytic colleague of mine recommended I read the diaries and letters of Etty Hillesum, someone I had not previously heard of. She knew of my interest in the nature of love and kindness, particularly as it has manifested itself in extreme situations such as the Holocaust. I make no apologies for the extensive reflections on Etty that follow, I believe that her capacity for loving may be an exemplar for how we could respond during the climate chaos to come.

Before Westerbork was established in the summer of 1942 Etty – single, intellectual, sensual, bohemian – had explored life to the full despite the Nazi occupation. She was sexually emancipated and was involved in simultaneous affairs with her rather unorthodox Jungian psychoanalyst and with her landlord. Her diaries also reveal her developing spirituality which took the form of a passionate and pantheistic identification with both the natural and the human world, a world she saw as the representation or embodiment of God. As her political consciousness developed so did her extraordinary capacity to face the truth, including truths about herself and her own weaknesses. This meant that, unlike many other Dutch Jews, she quickly shed any illusions she had about the nature of Nazism and became increasingly convinced of its genocidal intent.

Etty's diaries and letters[1] are both a warning and an inspiration. Her diary entries begin when she is still a free woman living in Nazi occupied Amsterdam. The Nazi dragnet eventually caught up with her in late February 1942. She reflects on her first encounter with the Gestapo, particularly one young irritable officer who had yelled at her,

I felt no indignation, rather a real compassion...he looked harassed and driven, sullen and weak.... I know that pitiful young men like that are dangerous as soon as they are set loose on mankind. But all the blame must be put on the system that uses such people. What needs eradicating is the evil in man, not man himself.[2]

The previous day Etty had been in earnest conversation with a friend who, in despair had asked what made human beings want to destroy others. Etty had insisted that 'the rottenness of others is in us, too'[3] and that owning this in herself and in her fellow Jews strengthened her capacity to love. Etty's faith in her fellow humans survived almost to the very end. She seemed to have had an almost infinite capacity to be understanding towards the weaknesses that many of her camp inmates displayed, particularly their desperate and at times pathetic hope. She also had a remarkable detachment and a forensic capacity to observe the many bizarre aspects of camp life such as the cabaret nights that prisoners were expected to attend and perform in whilst just outside trains awaited the next embarkation of passengers for Poland.

How did she keep on keeping on? It seems that her identification with nature played a crucial part. Here she is with Jopie, a fellow camp inmate:

> Now and then I join the gulls. In their movements through the great cloudy skies one suspects laws, eternal laws of another order than the laws we humans make. This afternoon Jopie, who feels thoroughly sick and all in, stood together with his sister-in-arms Etty for at least a quarter of an hour looking up at one of these black and silver birds as it moved among the massive deep-blue rain clouds. We suddenly felt a lot less oppressed.[4]

Etty seemed to be in touch with some great elemental flow in life, a connection that appeared never to have faltered. Here she is again,

> The misery here is quite terrible; and yet late at night when the day has slunk away into the depths behind me, I often walk with a spring in my step along the barbed wire. And then time and again, it soars straight from my heart – I can't help it, it's just the way it is, like some elementary force – the feeling that life is glorious and magnificent, and that one day we shall be building a whole new world. Against every new outrage and fresh horror, we shall put up one more piece of love and goodness.... We may suffer, but we must not succumb.[5]

Etty names this force that rises within her as love, and for her this love is the presence of God. As she puts it, 'love for everyone who may cross your path,

PARADISE LOST? The Climate Crisis and the Human Condition

love for everyone made in God's image, must rise above love for blood relations'.[6]

Etty's capacity to bear the suffering of those around her in Westerbork was remarkable. But it did affect her, not so much in the soul or mind, instead, she writes, for her 'the sufferer is the body'.[7] What can't be borne is projected, enacted or somatised and with Etty it was definitely the latter – she was repeatedly laid low by illness. Moreover there were moments when she did temporarily succumb, when she had to witness events which were too difficult for her to bear. Here she describes one such occasion and the mental dissociation that it wrought in her.

> But the babies, those tiny piercing screams of the babies, dragged from their cots in the middle of the night...I have to put it all down quickly, in a muddle, because if I leave it until later I probably won't be able to go on believing that it really happened. It is like a vison, and drifts further and further away. The babies were easily the worst.[8]

Her diary entry for June 8th 1943 is one of the most stunning and affecting pieces of writing I have come across anywhere. I am tempted to reproduce the two-page entry in full. She describes the 'enraptured expression' of a German guard as he picked purple lupins 'his gun dangling on his back'. There's a train 'billowing white smoke' to her left, its thirty-five freight cars already nearly fully loaded with Jews going to their deaths. There's a matron from the camp orphanage 'carrying a small child in her arms who also has to go, alone'. Layer upon layer of horror she describes until the engine gives a 'piercing shriek' and moves off with its cargo, and then she concludes with this:

> The sky is full of birds, the purple lupins stand up so regally and peacefully, two little old women have sat down on the box for a chat, the sun is shining on my face – and right before our eyes, mass murder. The whole thing is simply beyond comprehension.[9]

It is at that point that Etty's capacity, despite everything, to find beauty and meaning in life, reaches its limit. Then, to her horror later that summer she spots her own mother, father and brother Mischa arriving at the camp on a transit train. She described it as the worst day in her life and it forced her to reflect upon the nature of the love for humanity that had been her guide. Etty describes it thus,

> Love for one's fellow man is like an elemental glow that sustains you. The fellow man himself has hardly anything to do with it.

But this love leads her to a surprising conclusion. She hears that her mother and father are to leave on the train for Poland in a few days and she speculates that her brother may well volunteer to go on the same train 'out of love for his parents'. I found myself shocked by what she then adds,

> In the same way that Mischa wants to go out of love for his parents, I won't be going – out of a different kind of love. Perhaps it is a more cowardly love, but I feel myself strong. I believe it is easier to pray for people from a distance than to see them suffer right next to you.

And in her next letter Etty adds, '[I]t is not fear of Poland that keeps me from going along with my parents, but fear of seeing them suffer. And that too, is cowardice'. So there you have it, Etty seems to tell us that a spiritual love, a love for all humanity, is in some crucial ways an easier love than an earthy love; that it is easier to love what seems like an abstraction than to choose to suffer with those with whom one is intimately entangled.

So is this love 'for one's fellow man', as Etty put it, a more detached and abstract love? I find it difficult to be certain. On the one hand she herself contrasts it with Mischa's 'different kind of love'. But it is also possible that Etty's spiritual love *was* an embodied rather than detached one, that she loved humanity as if it *were* her family, as if the strangers she cared for were her neighbours or lovers.

Whatever the case, in choosing the love that she did Etty was able to bear witness and pass on the immeasurable gift of her testimony to future generations. Perhaps this is always what happens in times of acute crisis when politics literally becomes a matter of life and death. Some choose to follow what seems like a more abstract love and fight for their principles perhaps even against the needs or wishes of their families, and others choose not to, but to put their loved ones first like Mischa. Perhaps this helps explain the 'astonishing' obedience of many Jews that Vasily Grossman laments in his *Life and Fate*.[10]

Tragically, in the arbitrary way that camp decisions were made, at the very last minute Etty was instructed to leave Westerbork and travel on the same train as her family, they in wagon 1, Etty in wagon 12. Typically she manages to scribble a last message on a card which she threw out of the train, it was found by farmers outside the camp and posted by them. It is to her friend Christine. It begins thus,

> Opening the Bible at random I find this: 'The Lord is my high tower.' I am sitting on my rucksack in the middle of a full freight car. Father, Mother, and Mischa are a few cars away. In the end the departure came without

PARADISE LOST? The Climate Crisis and the Human Condition 159

warning…. We left the camp singing, Father and Mother firmly and calmly, Mischa too. We shall be travelling for three days.

There's a portrait photograph of Etty on the opening page of her diaries and letters. She is resting her chin on her upturned right hand, a cigarette grasped between two of its fingers. She is looking at us and her look is searching. She looks cool and tough and modern and has an extraordinary beauty.

Collection Jewish Museum, Amsterdam, Photo B. Meylink, 1937-8

Senseless kindness

In Vasily Grossman's great sprawling tragedy *Life and Fate* there is a famous passage[11] which purports to be the scribblings of Ikonnikov, a crazed mystic who has been imprisoned by the SS in occupied Russia. Ikonnikov speaks for Grossman, and this is one of the most impassioned parts of the book. Here is Grossman's paean to 'senseless kindness', a kindness which is powerful only when it is powerless, only when it remains a simple expression of ordinary goodness rather than a manifestation of what Ikonnikov calls Good with a capital G. For Ikonnikov / Grossman, all the worst crimes are committed for the sake of this Good, all the worst forms of human violence are committed out of love for this Good (which is of course an ideal good, one that can be embodied in a faith, ideology, country, ethnic group or imagined race).

This struggle between ordinary goodness and the Ideal is a recurring theme in psychoanalysis. Freud believed the group was in thrall to the Ideal, and in appropriating all goodness to itself it exported all badness into the Other, something Etty Hillesum refused to do. Moreover, the Ideal found representation in the leader who became the admired and feared super-ego, the big man, the Ubermench, the Ideal incarnate. In abandoning their conscience to such leaders the followers became capable of the most inhuman acts. According to Freud:

> In obedience to the new authority he may put his former conscience out of action, and so surrender to the attraction of the increased pleasure which is certainly obtained from the removal of inhibitions.[12]

Following Freud, Janine Chasseguet-Smirgel[13] traces the connection between the Ideal and ideology. She believed that all groups must have their ideology, all must be devoted to their own mythical Good.

Returning to Ikonnikov's musings, he says that the Good is not the enemy of Evil, no Good and Evil need each other, beckon each other, embrace each other. According to Ikonnikov, human history is not the struggle of Good against Evil but the struggle of evil to crush a small kernel of human kindness:

> It is not man who is impotent in the struggle against evil, but the power of evil that is impotent in the struggle against man. The powerlessness of kindness, of senseless kindness, is the secret of its immortality. It can never be conquered. The more stupid, the more senseless, the more helpless it may seem, the vaster it is.[14]

In her fascinating book *The Heart of Altruism,*[15] Kirsten Monroe reported her in-depth interviews with Germans who had rescued and/or sheltered Jews during the Holocaust. Over and again the same theme emerged from her interviewees, that the rescuers' actions were not guided by faith or ideology or any greater Good but were spontaneous reactions to the plight of the other – they had to do it, they had no choice. In my book *Politics Identity and Emotion* I explored the nature of this impulse:

> it is as if it were something that silently inhabits the person, only making itself present at certain key moments...this moral impulse almost

resembles a force that acts through the self – almost like a stranger, an external force, a force of necessity, even fate.[16] [hence perhaps part of the reason for the title of Grossman's book]

Alex Danchev[17] noted how influential Grossman was for the French spiritual philosopher Emmanuel Levinas, who draws from it the force of the face of the other's suffering, and the call it makes upon us.[18] And what about the non-human face? Is it meaningful to speak of nature's face and the call it makes upon us in its suffering? Is this the same as Etty Hillesum's love, a love for all of humanity, indeed perhaps a love for all sentient life?

This brings to mind work I conducted twenty years ago with a team researching the ethical dilemmas faced by youth and community workers who operated in some of Britain's poorest districts.[19] What struck us very powerfully was the compassion of these workers, their attentiveness and solidarity with young people who had not only been deeply damaged by life but who manifested this in their damaging behaviour towards others around them.[20] As one youth worker put it, 'they scratch at each other's pain'. It struck me that true compassion does not require the other to be good or innocent but is extended to the other in spite of their emotional ugliness and moral frailty. This is what 'the benefit of the doubt' means. This really is the true heart of generosity.

The many faces of goodness

You'll remember that what Etty Hillesum called 'the worst day of my life' was the day that her mother, father and brother arrived at the transit camp. She chose not to go with them on their final journey but as fate would have it she ended up on the same transport to Auschwitz. You might say that Etty Hillesum was guided by some higher form of love, in contrast to the earthy and intimate love felt by lovers, parents, siblings and, in some cultures, by kin. Yet clearly Etty felt the tension between these two forms of love and the contrary claims they made upon her. It wasn't that Etty was unloving towards her family it was just that her more spiritual love required a freedom from the earthly ties of family.

We have seen how this difference between private and public love also manifests itself in more ugly ways. Dan Gretton's work illustrates how kind and loving fathers, like Rudolf Hoess the commandant at Auschwitz, can be public monsters.[21] And it is equally possible for paragons of the public good to be private monsters.[22]

Interestingly enough many of Kirsten Monroe's rescuers had few illusions about themselves. Margot, a rescuer who undertook incredible risks – which included sleeping with a Gestapo commander in order to get information – simply said, 'I have my good sides and my disagreeable sides. I can be a bitch; I can be nice'.[23] Monroe's respondents, like Etty Hillesum, had no need to idealize themselves and they had no need to demonise the other – if the Nazis were monsters they were made of the same stuff as everyone else. Monroe adds, 'this appears to give many (but not all) altruists a tremendous understanding for weakness and human frailty and a remarkable forgiveness of even the vilest deeds'.[24] Perhaps the opposite of the altruist by this definition is the moralist. Tzvetan Todorov argues that moralism 'consists of practising justice without virtue, of simply invoking moral principles without applying them to oneself, of assuming one's own goodness simply on the basis of having declared adherence to principles of good'.[25]

Etty's love for all sentient life seems very different to the senseless kindness manifest by many of the 'rescuers' interviewed by Kirsten Monroe. Perhaps only if my love embraces the otherness of the other, only if my love for them is precisely because they are not-me or not-my-group, only then do we have goodness with a little g, ordinary goodness? Can this smaller good assert any political power or is it doomed to exist unseen at the margins of political life? Would the care shown by intensive care staff towards unvaccinated Covid victims qualify as this kind of unconditional love? And what about those in the climate movement who seem motivated primarily by a love for all living beings, human and nonhuman?

Ordinary ideals and the common good

It seems that we should be profoundly suspicious of any kind of group Ideal or ideology for it is in the name of such ideals (the motherland, Christianity, 'the people', etc.) that some of the worst crimes in recent history have been committed. Remember what was said in Section 2 of this book: whereas liberal democracy is afraid of the truth and authoritarianism is contemptuous of the truth, fundamentalists believe they have exclusive possession of the truth and therefore hold the key to the total explanation of all things (thus the term 'totalitarianism').

But there is a danger that a profound suspicion of ideology leads us to question the value of ideals per se. Surely ideals do not necessarily lead to idealisation and fundamentalism, for without ideals we have no measure of truth or value. Life without ordinary ideals would be a morally vacuous life where we have nothing to strive for, no measure of goodness against which to measure ourselves. Although, as we have seen, shame can be a very toxic emotion, we must not forget the importance of the healthy sense of shame which arises when one has fallen short of one's moral standards. To feel ashamed in such instances is to be able to say to oneself 'but I am better than that'. Healthy shame, like guilt, can provide a vital internal corrective, one which helps us to adjust our course in a way which is more congruent with our ideals. One only has to think of shameless figures such as Donald Trump or Boris Johnson to understand the moral value of shame.

Our ideals therefore seem central to our moral and ethical lives, something more than the spontaneous acts of senseless kindness that Grossman speaks of, something less than the Good in whose name so many acts of violence have been committed. I find myself being pulled towards the idea of the common good. 'Common' resonates with the 'commons', a reminder of a world which was held in common, predating its enclosure into private property. It also resonates with something ordinary rather than special or exceptional and therefore has a resistance to being hijacked by groups puffed up with passionate intensity. Whereas the Good always flirts with the Ideal, with fundamental truths, the common good is a fiction and all the more powerful for being so. I use the word 'fiction' here to indicate that the common good per se does not and cannot exist but this in no way undermines its importance as a fictional idea. The common good is an illusion, a play material of the mind, and as such populates the space in which politics takes place, a space where different goods compete and collide. The common good is an example of what in the previous chapter I called a 'mobilising fiction', a particular kind of fiction which has the power to inspire, move and galvanize political action.

In our struggle to combat climate change, to oppose fossil fuel interests and the finance capital that supports them, in our struggle to persuade governments to act both individually and collectively, we are motivated by a set of ideals about what constitutes the common good. Care constitutes one such ideal. A good society would be one that put care for both the human and the nonhuman at its centre. It would value interdependency, non-exploitative relations, respect for otherness and for the sentience of all living things.

The tension that Etty experienced between the claims of humanity and the claims of her immediate loved ones meant that she was living in what Bonnie Honig[26] calls 'dilemmatic space', the space of conflicting claims, loyalties and obligations, a space which challenges our moral and ethical capacities. In this space there is often no right thing to do, whatever choice one makes involves hurt or loss. As we have seen, this is the sphere of tragedy. We all live out the contradictions of the complex and diverse society in which we live on a day-to-day basis and, as a consequence, we are pulled this way and that. Honig draws on the work of the moral philosopher Bernard Williams who is keenly aware of the incommensurable nature of human values. Things just don't fit together as we would like them to, values rub up against each other, the moral agent has to live with conflicts that cannot easily be resolved and simply have to be lived with. Sometimes you have to end up disappointing someone. Williams argues that in such situations there is often no right thing to do, all we can do is 'act for the best'.[27]

---

[1] Hillesum, E. (1999) *An Interrupted Life: The Diaries and Letters of Etty Hillesum 1941-43*, pref. E. Hoffman, intr. J.G. Gaarlandt, trans A.J. Pomerans. London: Persephone Books.
[2] Ibid, p. 105.
[3] Ibid, p.103.
[4] Ibid, p. 366.
[5] Ibid, p. 355.
[6] Ibid, pp. 396-7.
[7] Ibid, p. 318.
[8] Ibid, p. 403.
[9] Ibid, p. 332.
[10] See my reflections on this in the previous chapter.
[11] Grossman, V. (2006) *Life and Fate*, trans Robert Chandler. London: Vintage; pp.388-395.
[12] Freud, S. (1921) Group Psychology and the Analysis of the Ego, Standard Edition of the Complete Psychological Works of Sigmund Freud, Vol. 18, p.85. London: Hogarth Press.
[13] Chasseguet-Smirgel, J. (1985) *The Ego Ideal: A Psychoanalytic Essay on the Malady of the Ideal*. London: Free Association Books. Chapter 4.
[14] Grossman, see note 11 above, p.394.
[15] Monroe, K. (1996) *The Heart of Altruism*. Princeton, NJ: Princeton University Press.
[16] Hoggett, P. (2009) Politics, Identity & Emotion. Boulder, Col: Paradigm. p.123.
[17] Danchev, A. (2013) Ethics after Auschwitz: Vasily Grossman and senseless kindness. *Journal of European Studies*, 43(4): pp.357-372.
[18] In her recent book Donna Orange offers an extended discussion of the relevance of the radical ethical position of Levinas to the climate emergency. Orange, D. (2017) *Climate Crisis, Psychoanalysis, and Radical Ethics*. London and New York: Routledge. Chapter 4.
[19] Hoggett, P., Mayo, M. and Miller, C. (2009) *The Dilemmas of Development Work: Ethical Challenges in Regeneration*. Bristol: Policy Press.
[20] Hoggett, P. (2006) Pity, Compassion, Solidarity. In Clarke, S., Hoggett, P. and Thompson, S. (Eds.) *Emotion, Politics & Society*. Basingstoke: Palgrave Macmillan.
[21] See my discussion of The Inhuman in Chapter 11. Hoess gave instructions to his children to always wash the strawberries that they picked in their beautiful garden, a house and garden coincidentally which was built right next to the walls of Auschwitz (ash and soot from the crematoria tended to form deposits on the garden's fruit).
[22] Patrick (Lord) Devlin, a celebrated British Judge who led public inquiries into the abuses of British colonialism and supported the appeal of the Guilford Four against their wrongful imprisonment, subjected his daughter Clare to years of sexual abuse.

Campbell, B. (2021) 'Our silence permits perpetrators to continue': One woman's fight to expose a father's abuse. 25th July, 2021. Available at https://www.theguardian.com/uk-news/2021/jul/25/our-silence-permits-perpetrators-to-continue-one-womans-fight-to-expose-a-fathers-abuse (accessed 6 September 2023).

[23] Monroe, see note 15 above, p. 208.

[24] Ibid., p. 207.

[25] Todorov, T. (1996) *Facing the Extreme: Moral Life in the Concentration Camp.* New York: Henry Holt & Co. pp. 114-5. Building on this principle Todorov adds that the mote in my own eye ought to bother me more the beam in my brother's. How unlike today's 'virtue signalling' one might be justified in thinking.

[26] Honig, B. (1996) Difference, dilemmas and the politics of home. In S. Benhabib (ed.) *Democracy and Difference: Contesting the Boundaries of the Political.* Princeton, NJ: Princeton University Press.

[27] Williams, B. (1973) *Problems of the Self: Philosophical Papers, 1956-72.* Cambridge: Cambridge University Press. p.173

CHAPTER TWENTY

ON NEW BEGINNINGS

The pulse of life

The tragic position that I described in chapter 17 seems to encourage a stoical attitude to life, to keep 'going on, going on' as the writer Samuel Beckett might have said. It is concerned with facing difficult truths (the strapline of the Climate Psychology Alliance) and with the 'pessimism of the intellect', believing that in doing so strength is to be gained. I am not referring to a brittle kind of toughness, but an emotional strength which enables us to take in more, hold and contain more of the complexity of life and thereby deepen our engagement with it.

This is the territory where psychoanalysis and existentialism have set out their stalls. But, as I mentioned in my discussion of hope, there is something missing, something vital. When my daughter was a few months old and crying in the middle of the night, I would sometimes take her downstairs and, as she rested her soft, warm head upon my shoulder, we would dance to *Music for Eighteen Musicians*.[1] Nearly always she would soon fall asleep, calmed not just by my holding of her but also by the rhythm. For me the pulse of life has always run through this music, the same pulse that fills me with joy when I first notice spring pushing up through the soil or when I get knocked over by the sheer vitality of a human toddler or by a group of starlings trading banter on a telegraph wire.

In early 2021 Extinction Rebellion UK put out an invitation to join in preparations for COP 26 via something they called the First Trimester project. The invitation asked, 'What if, in the months leading up to COP26, we could conceive of and "gestate" ideas of a new era'. Later in the invitation it quoted Jean Paul Lederach, someone who has spent his career in conflict transformation, speaking of *the moral imagination* which he defined as 'the capacity to imagine something rooted in the challenges of the real world yet capable of *giving birth* to that which does not yet exist' (my emphasis).

Natality

A connection immediately formed in my mind to the idea of 'natality', introduced by Hannah Arendt in her book *the Human Condition* back in 1958. This book was written in between two others which constitute the cornerstones of her work – *On Totalitarianism* in 1951 and *Eichmann in Jerusalem*

PARADISE LOST? The Climate Crisis and the Human Condition 167

in 1963. Arendt was clearly someone drawn to facing difficult truths, just as we are called upon to do the same today with the world facing the twin (and related) crises of runaway climate change and the return of authoritarianism. In spite of, or perhaps because of, the darkness that she looked into, Arendt's belief in the power of human goodness remained undaunted and it was this belief that was crystallized in her idea of natality. As one of Arendt's biographers Samantha Rose Hill noted, 'in the darkest hour of her life, as she contemplated suicide in an internment camp, she decided she loved life too much to give it up.'[2]

For Arendt the idea of natality, of new beginnings, constituted a much more reliable and powerful force for good than hope. She believed that natality and action were inextricably connected. According to Arendt,

> action has the closest connection to the human condition of natality; the *new beginning* inherent in birth can make itself felt in the world only because the newcomer possesses the capacity of beginning something anew, that is, of acting.[3]

Later she says,

> to act...means to take an initiative...to set something in motion...[4]

And adds that the:

> character of startling unexpectedness is inherent in all beginnings and in all origins.... The fact that man is capable of action means that the unexpected can be expected from him, that he is able to perform what is infinitely improbable. And this again is possible only because each man is unique, so that with each birth something uniquely comes into the world.[5]

Vitality

As I thought about this, a number of terms gathered in my mind – a will for life, joie de vivre, jouissance, a life or vital force, vitality.... We see this same will for life in our own young children and grandchildren and we still see it in children at play in the ruins of Syrian cities and Greek refugee camps.

So off I went to the website Psychoanalytic Electronic Publishing, the Wikipedia of the psychoanalytic world. Interestingly, when I did a search under the term *'vitality'* I got a massive 4734 results, that's over 4000 articles in which the term is used. It's ubiquitous. I also did a search under the heading 'will' and found a similar result. It follows that a vast number of psychoanalysts talk about vitality and 'will' but in a quite unreflexive, taken-for-granted, kind of way. This point was made quite forcefully by one of the few articles I found (which focussed on one of Freud's dissident colleagues Otto Rank) that makes the concept of 'will' an explicit focus for investigation.[6]

There is just one other interesting exception to this, my search under 'vitality' revealed that Daniel Stern's last book, published just two years before his death in 2012, is called *Forms of Vitality*. Stern, a developmental psychologist and psychoanalyst, insisted that the experience of vitality remained largely 'hidden in plain view', particularly within the therapeutic world itself. Stern saw vitality as something essentially embodied in movement. It was something in us which both moves and is moved by others, sometimes formally though music, theatre or art, but more usually through the style, form, dynamic and tone of our engagement with the world. Vitality is about movement and flow, it is analogic, unlike speech which is digital which comes in chunks, bites and units.

At first this stress upon 'movement' unnerved me somewhat until I began to think of 'being moved'. This is such a vital component in any genuine form of compassion, one which involves an active surrendering to the moving presence of the other. And then the political self in me started firing: I suddenly remembered how fundamental the idea of movement is in politics, the centrality of political and social movements to our thinking about change. And then, via another association, to the idea of society as a balance of forces, some pushing for change some resisting it, sometimes leading to movement and sometimes (as with Gramsci's concept of 'the interregnum') leading to stasis and paralysis. And how like this the mind is too! So movement, forces pushing in different directions, change occurring in space and time – Stern sees these elements (movement, time, force, space, directionality) as the Gestalt comprising vitality. He offers a list of words which throw

PARADISE LOST? The Climate Crisis and the Human Condition   169

light on vitality. One of them is 'pulsing' and I immediately think of the pulse of life in *Music for Eighteen Musicians*.

Articles influenced by Stern's book are now appearing in the various professional journals. I noticed that one of the very first was in the Infant Mental Health Journal, no coincidence given Stern built his reputation upon detailed studies of mother/infant interactions. The authors note the crucial role of these early interactions in establishing a secure sense of agency and efficacy (the experience of having one's agency confirmed) in the developing infant.[7] They add that this efficacy resides in the perception that through one's action one has changed and can change a course of events.

How interesting! We have come full circle from Hannah Arendt's idea of natality and its central place in her philosophy of action. Remember what I cited from her earlier,

> 'to act...means to take an initiative...to set something in motion'
> the 'character of startling unexpectedness is inherent in all beginnings'.

We are in the vicinity of some kind of force which is crucial to being alive, a force which in some essential way is physical, bodily and earthly. Whereas the tragic position sees in our nature only our frailty, transience and mortality, from the perspective I am offering here the nature within us is also the medium for our liveliness, spirit and creativity. This medium is a trans-subjective one, shared with all living things and pre-existing the individuated self. In the infant it is manifest in the joy of exploration, of reaching out, of being known, of being curious, of wonder and surprise. And in rare moments, which are worth waiting for, it inhabits the life of politics as well – unafraid, riotous, carnivalesque, defiant.

This is the force that Arendt – in her internment camp – was in touch with. Arendt had faced loss, ruin and mortality and in the process had 'found' natality, a new beginning. This is the task that faces us societally and politically right now, to summon our strength, humanity and moral imagination to face up to and oppose the inhuman in society and in ourselves. Amitav Ghosh pictures this in terms of the emergence of a 'vitalist politics', one which grounds itself in the vitality of Earth's natural forces, a vitality that the mechanistic philosophies of modernity sought constantly to suppress.[8]

A love for life

I want to return to the thoughts of Arendt and Grossman about the Warsaw Ghetto. They argued that hope disarms us, makes us obedient and compliant and it does this because of the fear that we feel for the survival of ourselves and our loved ones. A few years ago I consulted with a team who provided therapeutic support to those with terminal or life-threatening illnesses. We were thinking about the sources of strength people found when faced with the uncertainty of death (for example, not knowing whether you have two months or twelve months to live), when one of the team mentioned the opposite, how some people would humiliate and embarrass themselves in their desperation to live at any cost.

In my examination of hope I argued that in facing the worst we become stronger. Now I would add that specifically it is our *will* that becomes stronger. I believe that I am talking about a will for life, a fervent belief in the value of all living beings, a yes to life. *This will for life is absolutely not the same as the will to live*, that desperate clinging to life no matter what the cost to dignity, beauty or love. Etty Hillesum, writing from the Westerbork Transit Camp in 1942, makes this difference touchingly clear:

> If we were to save only our bodies and nothing more from the camps all over the world, that would not be enough. What matters is not whether we preserve our lives at any cost, but *how* we preserve them.[9]

Overcome by terror, the survivalist individual or group becomes de-moralised and questions of value are rendered meaningless, destroyed. The survivalist has lost their love of the world, instead of a secure attachment to life there is a desperate clinging. This can lead to the fatalism, passivity and obedience that Arendt and Grossman observed but it can equally lead to ruthlessness and cruelty. Faced with social catastrophe the survivalist group does what it must do in order to survive. In the landscape of Cormac McCarthy's novel *The Road*, this is a vigilant and predatory group whose brutality is fired not so much by sadism but by a cold resolve to live. Where the choices are seen as victim or predator, it is better to prey than be preyed upon. What distinguishes human beings from animals steadily disappears. It is a survival of the fittest.

'From gaining the sympathies of a Gestapo guard through her gift for storytelling, to having the courage to walk out of an internment camp with forged papers, to journeying across France alone on foot in search of her husband, Hannah Arendt acted courageously time and again'.[10] Arendt practised what she preached. Far from a passive and compliant clinging on to hope or demoralised survivalism *she had found a love for life that was stronger than a desperation to live.*

[1] This piece of music by the north American composer Steve Reich, first performed by his band in 1976, was a key early example of what came to be known as minimalism in music. Interestingly it consists of 11 sections which Reich actually calls 'pulses'. Probably more so than any other minimalist composition its influence crossed over into the dance and electronica fields in subsequent decades and their use of loops, layering and phasing.

[2] Samantha Rose Hill (2021) When hope is a hindrance. AEON, 4 October 2021. Available at: https://aeon.co/essays/for-arendt-hope-in-dark-times-is-no-match-for-action (accessed 6 September 2023).

[3] Arendt, H. (1998 (1958)) *The Human Condition*. Chicago: University of Chicago Press. p.9.

[4] Ibid, p.177.

[5] Ibid, p.178.

[6] Lieberman, E.J. (2012) Rankian Will. *American Journal of Psychoanalysis*, 72: 320-325. Interestingly enough Paul James, whose criticism of Martha Nussbaum's concept of human capabilities I will mention in Chapter 23, also sees vitality as an absolutely key aspect of human flourishing.

[7] Ammanaiti, M. and Ferrari, P. (2013) Vitality affects in Daniel Stern's thinking: A psychological and neurobiological perspective. *Infant Mental Health Journal*, 34(5): 367-375.

[8] Ghosh, A. (2021) *The Nutmeg's Curse: Parables for a Planet in Crisis*. London: John Murray.

[9] Hillesum, E. (1999) *An Interrupted Life: The Diaries and Letters of Etty Hillesum 1941-43*, pref. E. Hoffman, intr. J.G. Gaarlandt, trans A.J. Pomerans. London: Persephone Books. p.305 (emphasis in original).

[10] See Hill, When hope is a hinderance, note 2 above.

CHAPTER TWENTY-ONE

REVISTING AGENCY

The agent and the reagent

In Chapter 2 I examined the way in which modern Western society conceives of the self as an active 'doer', shaping its own life[1] and the world all around it. I also noted how this way of seeing could only be sustained by suppressing two equally important dimensions of reality – where nature was an active co-determining subject, and where the self was an object of forces beyond its comprehension or control.

This idea of the self as the creator, sometimes referred to as the 'active voice' of the self, was reinforced by the rationality of property which became increasingly entrenched as capitalism consolidated its hold, first in Europe, then the rest of the world. One consequence of this rationality is that the impulse to take ownership reaches deep inside the modern self where it connects to that primordial longing to be in possession of the Ideal.

Etty Hillesum captures this when, in her diaries, she reflected on her urge to write.

> I think I know what all the 'writing' was about as well: it was just another way of owning, of drawing things more tightly to oneself with words and images. And I'm sure that that used to be the very essence of my urge to write: I wanted to creep silently away from everyone with all my carefully hoarded treasure, to write it all down, keep tight hold of it, and have it all to myself.[2]

Ownership also operates within the group psyche. Take the idea of 'climate psychology'. The Climate Psychology Alliance staked an early claim to this concept, wrote books using the term, wrote an entry in Wikipedia which was careful to ensure that 'our' take on the concept was included at the inception of its voyage. But if we hadn't come up with the idea then someone else would have; the time was 'ripe' for it. The conception had already occurred, what was required was a midwife for the new idea to be born.[3] When I say here 'the time was already ripe' I am referring to the eco-psycho-social system that we are all tiny elements within. This alternative way of thinking about agency challenges the very proposition that 'we' had an idea, that the thinker preceded the thought. Bion always insisted that the thought precedes the thinker, the thinker is the vessel through which the thought finds expression.

Of course we'd like to think of ourselves as agents who created this thing called 'climate psychology'. But this is a modernist conceit. We just happened to be at the right place in the right time. Not agents but vessels, catalysts rather than creators. I like what this does to our thinking about agency. Rather than the active hubristic voice of modernity and individualism we find the passive voice, rather than control and command we have surrender, letting go and letting come.[4] Rather than 'making things happen' we have 'generating the conditions' (in which something new might emerge). Rather than goal-oriented, task focussed planning we have something more improvisational and transgressive – 'Say we did X, let's see what then happens'. This is experimental, but not in the positivist sense of the controlled experiment, more in the premodern sense of an alchemical process in which the individual or group has the status of a (re-)agent, an element which helps bring about system transformation. I believe that it is this model of reagency that CPA shares in common with the new wave of social movements including Extinction Rebellion. I would like to think of the CPA as a work of the commons where we all dip from the same common pool for new ideas and practices without claiming property rights over them.[5]

By questioning the modern idea of agency I believe we can begin to rethink 'the commons'. In Chapter 2 I mentioned the destruction of the commons which accompanied the appearance and development of private property and capitalism. In thinking of the commons in this way our mind is taken to physical things, land in particular. Even as I speak subtropical rainforests which belong to no-one are being appropriated by capital, destroyed and then transformed into palm oil plantations or beef ranches. But say we thought of the commons in non-material terms, not as resources to be exploited but as shared ideas and practices. As such, the commons are gifts to be enjoyed within and between cultures, parts of complex, emergent systems momentarily manifest in innovations in thought, art, politics, policy or ethics.

Viewed from this perspective, public services such as free health care or education are a work of the commons,[6] as are shared practices such as Sufism or Mindfulness, cultural innovations such as Electronic Dance Music and social innovations such as participatory democracy or climate psychology. Then perhaps we can begin to see that 'the commons' is alive and well, indeed with globalism it appears to be burgeoning as the world becomes our shared oyster. I think that there is a connection here with Hannah Arendt's concept of natality, of giving birth to something new, of a life force at work in both the more than human and the human worlds.

The trans-subjective

I opened this chapter by noting how the modern self is conceived as a subject in a world of objects. This way of thinking about the self is to some extent shared by psychoanalysis where the idea of 'the object' is absolutely fundamental. Psychoanalysis speaks constantly of love objects, object loss, sexual object, object relations, choice of object, part objects, bizarre objects, and so on. Such objects are not real things but ways of representing experience and thus the basis of psychic reality. But perhaps if we are to really disrupt modernist ways of thinking about the self then we need to go further by de-objectifying our world.

In common with most mammals, human life comes into being in an inter-uterine environment. As we saw in Chapter 14, Michael Balint noted that the amniotic sea and the foetus, the container and the contained, exist together in a harmonious mix-up – yes the foetus is inside the mother, but the mother is also inside the foetus. At first 'communication' between the two is primarily biochemical, mediated by the mass of blood vessels of the placenta. But as the foetus develops so sensory capacities also form, first touch, then sound.

Reflecting on the work of Peter Sloterdijk, the German philosopher, Sacha Rashof[7] describes how he speaks of pre-objects or 'nobjects' which she suggests can be thought of as 'small interior comings-towards-the-world, which cannot conceive of themselves as objects since they have no other/s to oppose yet'.[8] Sloterdijk draws attention to the pre-oral phase of development, to the period of inter-uterine life. He sees this uterine environment as the earliest form of containment, an original 'micro-uteri' which forms the template for the multitude of macro-uteri which then populate the social world stretching all the way from the computer screen to the built environment and to spaceship Earth.[9]

Perhaps there are two layers of human selfhood, one relatively undifferentiated and trans-subjective, the other differentiated into subjects and objects, self and other. Some writers use the term 'matrixial' to refer to the former.[10] The origins of the word matrix[11] takes us back to Latin and Old French where it denotes the uterus and the womb, a place where new life forms and develops. This is not at first a space of hard boundaries and separations, of subjects and objects. The womb is a trans-subjective space where messages are exchanged long before the acquisition of words. They are transmitted rather than communicated, trans-subjective experience is analogic not digital. We must remind ourselves that analogic communication lies at the heart of psychodynamic practice – free association, the transference and countertransference, contagion, basic assumption mentality, all of this is

PARADISE LOST? The Climate Crisis and the Human Condition 175

primarily based upon flows of imagery and affect in which messages are transmitted but beyond conscious human awareness. For those who are into it, this is right brain stuff.[12]

Michael Balint, Daniel Stern and many of those within the therapeutic world who have been closely involved in observing parent/infant interaction draw our attention to the vital forms of communication, this world without words, which constitutes the foundation of human subjectivity. The trans-subjective therefore pre-exists the individuated self, constituting the ground or medium upon which individuation, and the idea of the individual mind, forms. The trans-subjective is organic, it pulses with life and affect. Whether we are aware of it or not we are immersed in this medium along with all living beings. In this realm there are no separate bodies interacting with each other but rather inextricably entangled beings engaged in intra-actions.[13]

I sit with my big black labrador on the cliffs near Hartland Point looking out across the Atlantic. The sun is out, there is a slight breeze, it is the early summer. She sits upright, enveloped by the sensory world, taking it all in – feeling it, listening to it, smelling it. We could sit like this for ages, sometimes we do, attending to it, surrendering to it, taking it all in.

Should we abandon the idea of agency altogether?

In Chapter 2 I explored nature as a sentient subject rather than passive object, as something that speaks to us if only we have the ears to listen. But what if nature doesn't just *speak to us* but also *acts upon us* and engages us actively in the shaping of our world? This is a perspective offered by the famous sociologist of science, Bruno Latour, who helped to develop Actor Network Theory (ANT) which sought to transcend the old dualisms of modernity, dualisms such as nature/society, or subject/object. Proponents of this approach argue that we should abandon these unhelpful binary ways of thinking and get on instead with tackling the world of actualities. ANT hence argues that all manner of things (as many as you can imagine) are variously entangled together in specific formations or networks in the making of the world.

This way of thinking which is sometimes called the *new materialism* tends not to speak of nature on the one hand and society on the other but of matter (which includes both). Nor does it talk in terms of human agents but of actors or actants which might be human but might not be. For example, the collapse of the West Antarctic Icesheet would have an impact on humanity far, far greater than the rise of Nazism and yet such a collapse, if it occurred, would be no more than a single event in a vastly complex emergent Earth system in which the promptings of humanity – a humanity which itself has become a force of nature – can be detected going back centuries. Perhaps then agency isn't something uniquely human. This is an anthropocentric presumption. The reality is that dogs, cows, starlings, trees, rivers and bacteria (the list is endless) all have agency. We need to put humans firmly back in their place.

It is a fundamental tenet of the kind of psycho-social perspective that I hold to that, far from always being conscious and intentional agents, we humans typically don't know why we do what we do at the time of doing it. This is partly because we are creatures of unthinking routine and habit but it is also partly because of the way in which we are often guided by *internal* forces of which we are quite unaware . For example, some people are unable to enjoy good fortune without some 'accident' swiftly following which destroys the new happy state of affairs. In my experience this often affects women who develop with what I think of as a 'narcissism deficit'. Having grown up to believe not to expect too much in life, to always put others first, they lack a healthy sense of entitlement. When good things happen in their lives, an unconscious sense of guilt is triggered which generates the phantasy that they will have to pay for this. The psychoanalyst Ronald Fairbairn referred to this as the 'internal saboteur'.[14] The 'accident' which often appears to come out of the blue and may, for example, take the form of the sudden onset of an apparently chronic illness, is the payback.

Following Latour and others, we might say that in such situations humans are actants (they still make things happen) but not agents. In a similar way, the trauma of one generation may be revisited on a future generation – it is not that it is done intentionally but its effects can nevertheless be profoundly destructive. And of course it is not just internal forces that we may be unaware of but external ones too. Indeed, one of the reasons I am writing this book at this time is because of my fear that citizens of western democracies are dangerously unaware of the way in which inaction around climate change is a crucial contributor to our slide towards nationalism and authoritarianism, developments which are likely to make action around climate change even more difficult. At times like these we are indeed born along like flotsam on vast currents beyond our understanding – certainly not agents, perhaps partly actants contributing to our own demise, but also objects of forces beyond our comprehension or our control. And these forces are not just an abstraction – 'the system' or 'the machine' – they are real, tangible things, existing in company headquarters, digging mines and oil wells, generating profits, spreading disinformation about climate change and sometimes no doubt laughing all the way to the bank.

Whilst I am sympathetic to the idea of the actant, there is one big problem with it. If humans are just one among many actants in a complex emergent system, how come the earth is threatened with destruction by us and us alone? Moreover, if we ditch the modernist idea of the individual agent, do we not also have to ditch the idea of the moral responsibility of the individual or group for their actions? This is the objection made by the Swedish ecologist and activist Andreas Malm[15] among others. He insists that by dissolving the boundary between the human and the nonhuman we let the former get away with murder. I agree.

So we must refrain from dumping the idea of human agency altogether. Yes, many actions are undertaken on the basis of conscious and strategic intent but many are also undertaken on impulse, as if some force takes hold of us that we are simply the vessels for. This is a different kind of agency, one I have likened to reagency. What I have tried to do in this chapter is to highlight the way in which agency has been infected by the individualism, the rationality and the fetish for ownership which is so much a part of modern society.

Aggression and the life force

Neither nature nor society comprise inert matter to be shaped and controlled by agents of change. Both are complex, ever evolving systems in which states of temporary equilibrium give way to new currents and forces. Into this medium human actors, individually and collectively, immerse themselves. So let's go back to the very first human medium, the amniotic sea. The psychoanalyst and paediatrician Donald Winnicott, thinking of the baby pushing at the lining of the mother's womb, said: '(I)n health the foetal impulses bring about a discovery of the environment, this latter being the opposition that is met through movement, and sensed during movement'.[16] The foetal impulse is therefore one of the very first signs of agency, synonymous with a vitality or life force which invigorates this tiny being as it pushes against the constraints and limits that contain it. Here then we have the original template of the action space or room for manoeuvre, something Winnicott called 'potential space'. Only by testing the limits of a situation can we discover it's actual, as opposed to assumed, boundaries. This is not a passive process of contemplation but an active process of experimentation which harnesses the aggressive energy inherent to the lifeforce.[17]

Love and aggression – perhaps these are the two basic affects comprising the life force.[18] And here I guess I may disagree with some of my colleagues in the climate movement who tend to believe that the movement should be all about love and peaceful direct action and who tend to think of aggression in a negative way. Andreas Malm has challenged the way in which non-violent direct action (NVDA) has been construed as the only legitimate form of protest available to our movement. After almost two decades of this – mass protests, school strikes, sit ins, glue ins, street occupations – Malm asks at what point do we escalate, when do we conclude that the time has come to also try something different, something that really does disrupt 'business as usual'?[19] He takes the Suffragette movement as his template. The Suffragettes were not at all afraid of attacking private property (mostly belonging to wealthy and/or powerful men) when this was necessary. They saw this as just another tactic alongside indirect action (lobbying, protest) and direct action (chaining themselves to railings, self-starvation).

Optimism of the will

The Italian communist Antonio Gramsci will be long remembered on the left for his maxim, 'pessimism of the intellect, optimism of the will'. By this he meant that real agency was built upon the courage to face reality, no matter how difficult this was. It required an unflinching critical gaze, 'directing one's attention violently towards the present', as Gramsci put it.[20] This is what he meant by 'pessimism of the intellect'. Far from leading to despair and the collapse of the will, in the right conditions facing the truth made the individual or group stronger, more determined and more imbued with a sense of possibility. This is what he meant by 'optimism of the will'.

The great paradox of modern civilization is that whilst it conceives of the self as a subject constantly acting upon the world, the reality is of an economic system, capitalism, with a malignant logic (growth) and self-destructive addiction (fossil fuels) which is out of control. As Stephen Duguid puts it, '(W)e do not believe in progress, we just do it....we have no way to get off'.[21] For the vast mass of Moderns, perhaps even the super-rich, there has been a collapse of real agency. One of the things that has replaced it is a magical belief in the possibility of personal transformation – the makeover, the upgrade, the 'high'. In this way the fantasy of the Ideal is maintained. This stands Gramsci's maxim on its head. Now there is an optimism of the intellect (a belief in fairy tales, wishful thinking) masking an actual fatalism and hopelessness, that is, a pessimism of the will.

---

[1] Naturally the psychotherapies have themselves been incorporated into this outlook. In particular there is a piece of psychobabble which is so ubiquitous at the moment that it drives me to distraction. I speak of 'the journey' and the metaphor of the self's journey.

[2] Hillesum, E. (1999) *An Interrupted Life: The Diaries and Letters of Etty Hillesum 1941-43*, pref. E. Hoffman, intr. J.G. Gaarlandt, trans A.J. Pomerans. London: Persephone Books. pp.18-19.

[3] In these passages I am drawing in my own way on the elusive but thought-provoking picture of thinking developed by Bion. See Bion, W.R. (1962) *Learning from Experience*. London: Heinemann.

[4] Here I would like to acknowledge a way of thinking developed within CPA by Chris Robertson who succeeded me as Chair of the organisation.

[5] Inevitably, given the hyper-individualism of our culture, we all struggle with this at times, and this can have a corrosive effect on our movements.

[6] Richard Titmuss who more than any other provided the philosophical foundation for the British Welfare State after World War Two used the metaphor of the blood transfusion service to evoke the idea of the common pool of humanity and the gift relationship underlying it. Titmuss, R. (1971) *The Gift Relationship*. London: Pantheon Books.

[7] Rashof, S. (2018) Spheres: Towards a techno-social ontology of place/s. *Theory, Culture and Society*, 35(6): 131-152.

[8] Ibid. p.133.

[9] Interestingly, Sloterdijk's thinking, particularly the idea that to define humans is to define their life support systems, is beginning to inform the work of some climate scientists. See Skrydstrup, M., Geissler, P., Sismnodo, S. and Kelly, A.H. (2016) Of Spheres and squares: Can Sloterdijk help us rethink the architecture of climate science? *Social Studies of Science*, 46(6): 854-876.

[10] In her essay 'How the light gets in' Wendy Hollway adopts this term from the artist and psychoanalyst Bracha Ettinger. Hollway, W. (2022) How the light gets in: Beyond psychology's Modern individual. In Hollway, W., Hoggett, P., Robertson, C. and Weintrobe, S. *Climate Psychology: A Matter of Life and Death*. Bicester: Phoenix Publishing House.

[11] The idea of the matrix is familiar to the Tavistock Group Relations tradition. One of its innovations, associated with the work of Gordon Lawrence, is called the social dreaming matrix. Lawrence had a background in social anthropology and believed that

dreams have a social as well as individual character. The social dreaming matrix was a methodology for the sharing of dreams and exploring the cultural referents within them. For an introduction to this approach see Manley, J. (2018) *Social Dreaming: Associative Thinking and Intensities of Affect*. Basingstoke: Palgrave Macmillan.

[12] McGilchrist, I. (2018) *The Master and his Emissary*. New Haven: Yale University Press.

[13] Barad, K. (2007) *Meeting the Universe Halfway: Quantum Physics and the Entanglement of Matter and Meaning*. Durham, NC: Duke University Press.

[14] Fairbairn, R. (1952) *Psychoanalytic Studies of the Personality*. London: Routledge & Kegan Paul.

[15] Malm, A. (2020) *The Progress of this Storm: Nature and Society in a Warming World*. London: Verso.

[16] Winnicott, D.W. (1950) Aggression in relation to emotional development. In Winnicott, *Through Paediatrics to Psychoanalysis*. London: Hogarth Press. 1975.

[17] Here I am drawing on some of the thinking I did about human agency back in Chapter 7 of my earlier book *Partisans in an Uncertain World*.

[18] And so, one might ask, what affects comprise that anti-life force that Bion hypothesises and which I have explored in terms of a 'laughing, ecstatic destruction'? David Bell thinks of this in terms of a Mephistophelian force deriving cruel pleasure in destruction and devoted to nullifying 'Nature's power to create'. This is hatred in its most elemental form, that is, hatred of life itself. Bell, D. (2015) The death drive: Phenomenological perspectives in contemporary Kleinian theory. *International Journal of Psychoanalysis*, 96: 411-423.

[19] Malm, A. (2021) *How to Blow Up a Pipeline*. London: Verso. Of course there is a cruel irony in the title of this book, for the year after its publication the massive Nord Stream pipeline was comprehensively sabotaged, perhaps permanently, by its effective owner Vladimir Putin. This led to massive reductions in energy consumption throughout Europe and gave significant impetus both to the switch to renewables and to nationalist strategies to enhance energy security. History moves in mysterious ways!

[20] Gramsci, A. (1971) *Selections from the Prison Notebooks*. London: Lawrence & Wishart, p.175.

[21] Duguid, S. (2010) *Nature in Modernity: Servant, Citizen, Queen or Comrade*. New York: Peter Lang Press

EXCURSION

BOWERCHALKE

Bowerchalke is a tiny village occupying the upper reaches of a short valley that cuts into the Cranborne Chase. There's something about the landscape of this Wiltshire/Dorset borderland that pulls me to it. I love the folded hills that roll from east to west. In places they are quite bare, covered only by chalk grassland or vast fields of wheat and other cereals. They curve and undulate like bodies partly immersed in the earth.

In many ways Bowerchalke is unremarkable. The chalk stream[1] that runs through it is barely visible in the overgrowth of the summer, the village itself is strung out along the length of the minor road that climbs south over the Chase to the next village which goes by the wonderful name of Sixpenny Handley. Like many of these rural villages the pub and post office and shop closed years ago and are now private residences. But the church remains and contains the burial place of the novelist William Golding, known for his Nobel Prize winning book Lord of the Flies.

Photo courtesy of Roger Lane[2]

Golding came to live in Bowerchalke in 1957, soon after the publication of his famous novel. By all accounts he was a troubled man, full of self loathing and very sensitive to criticism. He had a terrible inferiority complex particularly in relation to social class. He tended to self-medicate with alcohol, a habit that intensified when his novel The Spire, *published in 1964, received bad reviews. Golding's incipient depression deepened further when he lost his beloved sailing vessel Tenace in a collision in the English Channel in 1967.*

The same year a new neighbour arrived in the village by the name of James Lovelock, who by this time was working for the National Space Agency (NASA) in the USA. The two men became walking and drinking companions. According to Lovelock in either 1968 or 1969 he put his hypothesis about the Earth being a complex self-regulating system to Golding as they were walking to the post office. Golding who had a background in the classics suggested that Lovelock should consider calling it the Gaia hypothesis. The two men continued walking as they got into an argument, one they eventually realised had been caused by a misunderstanding – Lovelock having heard 'gyre' (a vortex) and not Gaia! Gaia was, for the ancient Greeks, the Goddess who personified the Earth.

In 1974, whilst still living in Bowerchalke, Lovelock wrote his famous scientific paper with Lynn Margulis which introduced the Gaia hypothesis to the scientific community. As Lovelock said about his hypothesis: '(U)nless we see the Earth as a planet that behaves as if it were alive ...we will lack the will to change our way of life and to understand that we have made it our greatest enemy'.[3] *I like to think that somehow the sensuous landscape of Bowerchalke and Cranborne Chase had an unconscious influence on Lovelock, rendering him open to Golding's suggestion.*

Meanwhile Golding's personal crisis had deepened. He began to suffer from writer's block and insomnia, he was besieged by family anxieties and sank into heavy drinking. But there was one area of his life which paradoxically began to flourish, as he put it to a colleague: 'he couldn't write but had these amazing dreams'. By the early 1970s, fascinated by the work of Carl Jung, he initiated a dream diary, one that he continued right up until the day before his death in 1993. The dream journal eventually totalled 2.4 million words!

In August 1971 whilst in Italy, Golding had a dream that would change his life. He would later refer to this as his 'great dream' and it seemed to initiate a process in the next decade which enabled him to emerge from his crisis and renew his creative writing. The dream is examined in detail in an illuminating essay by Tim Kendall.[4] *Suffice to say the dream is set on the Spanish Steps in Rome and the key figure is a very old man. According to Golding's dream diary,*

There was a very, very old man there who was related as closely to me as the steps were. He had been famous as a great singer and/or maker of folk songs, ballads and the like. He was Yeats, perhaps, or perhaps no one. He explained that he was too old now to sing; and he began to go away down some steep narrow steps at the side of the violin-shape. He became older as he went and more crooked; but as he went down into the darkness he began to sing. Immediately the whole mass of the Poetry Party rushed to the railings by the steps to listen, as the great voice went away into the dark.

Kendall suggests that the dream offers a vision of a creativity that goes singing down into the dark. As Golding writes in his journal: 'to look forward down the small slope to death is proper; and to find the work that should go with that look forward is proper' (13 January 1972). The work he eventually 'finds' consists of six further novels all, according to Kendall, bearing traces of the 'great dream' and one of which, Rites of Passage, *won the Booker Prize. Golding's great dream seemed to unlock what he referred to in his journal as his 'crisis'.*

Kendall reflects that Golding was 'fascinated by dreams because they make us all visionaries, receiving unbidden wonders from mysterious forces that leave us uncertain as to how (or whether) to act on them'. These 'mysterious forces' haunt both our internal and external landscape and are part of what in the last chapter I referred to as the trans-subjective. Golding and Lovelock in their different ways seemed able to surrender to these forces and, through their re-agency, then let something wholly new come into being.

---

[1] 86% of the Earth's chalk streams are to be found in England. They constitute a rare and fragile ecology.
[2] Available at www.rogerlanephotography.com. Reproduced from his book *'Cranborne Chase - A Secret Landscape'* published by Amberley Publications.
[3] Lovelock, J. (2007) *The Revenge of Gaia*. London: Penguin. pp.21-2.
[4] Kendall, T. (2018) William Golding's Great Dream. *Essays in Criticism*, 68(4): 466-487.

CHAPTER TWENTY-TWO

SUSTAINABLE ACTIVISM

Politics and the emotions

It sometimes seems that to avoid dangerous climate change in a just and equitable way, with our carbon budget already largely used up, requires a degree of cultural change which is simply beyond our reach. But perhaps this is to ignore just how much, in other areas, the culture of liberal democracies has actually changed in recent decades. A good example is the transformation achieved in the position of gays and lesbians in my own country, the UK, in the last 40 years. Could not environmentally destructive consumerism go the same way as homophobia?

Deborah Gould studied the rise and decline of the direct action AIDS movement ACT UP in the USA in the 1980s and early 1990s.[1] This was a movement built upon gay and lesbian communities, and Gould noted the powerful role that shame, pride and despair played in this activism. Gould's research was a direct challenge to the male dominated world of research on social and political movements which was, and still is, in thrall to the same rationalist assumptions that dominate economics. Only in the last twenty years have a small minority of researchers, including Gould, begun to take the role of the emotions in politics seriously.[2]

As Gould noted, any political movement seeking to make things better in the world has to manage despair. In politics, despair arises because from the beginning activists are haunted by the belief that they might lack the collective resources to address the damage and suffering that they see around them, and that motivates their action. In other words, besides its external opponent, a political movement always has an internal one, and that is despair.

Containing despair

In the period 2014-2015 my colleague Ro Randall and I interviewed activists involved in the first wave of direct action around climate change in the UK,[3] a wave which ended in 2009, and involved interventions such as the occupation of coal fired power stations and airport runways. We wished to explore how they managed the powerful feelings aroused by exposure to the disturbing truth of climate change. Here's one young female activist speaking about this despair:

I know if I let open the floodgates it's there…I know what that depressive, overwhelming 'I feel lost' feeling is. I've had it. It's not something I enjoy.

In our own experience of movements for change we have been struck by the way in which the failure to contain despair can lead to an unrealistic hope. The group 'puffs itself up' to make itself feel big; it overestimates its own strength and underestimates the power of the opposing forces. It resorts to faith ('history is on our side') and magic ('come on everybody, one last push'). It prefers to engage in wishful thinking than face reality as it is.

Failure to contain despair can also produce a state of mind, sometimes referred to as schizoid, that finds it hard to avoid thinking other than in terms of polarities. This state of mind is the outcome of probably the most fundamental defensive manoeuvre that we Moderns engage in, more fundamental even than repression. Besieged by internal anxiety we split and divide the world, as if a less complex world, one more black and white, would be easier to live with. We allow our world to become fractured around a whole set of oppositions – either/or, all/nothing, right/wrong, good/bad, thought/feeling, and so on.[4]

Avoiding binaries

The state of mind is one we often encounter in our work as a psychotherapists. This is the familiar world of binary thinking where it seems there is no 'in between'. Take either/or for example. Everything is either one thing or the other and the coin constantly flips between one perspective and its polar opposite – either my marriage was the wonderful relationship I always imagined it to be or I was living a total illusion; either I have this special (and exclusive) relationship with my children or I mean nothing to them. And the odd thing about this splitting is that it creates an ideal state of affairs that can never be achieved in reality. Failure, disappointment and rejection is inevitable and so splitting paradoxically reproduces the very anxiety (that I am a worthless nothing) it tries to manage.

Either/or thinking can affect the culture of otherwise healthy movements. Within movements around climate change we can see this at work in terms of a series of unhelpful binaries. For example, 'The only realistic thing to do is change the system' *versus* 'We are powerless to change the system we have to focus on achievable changes in our community and in our own lives' or 'You can never change the state you have to fight against it' *versus* 'There is no alternative to working within the state for change, anything else is unrealistic'.[5]

Another binary is 'all or nothing'. We throw ourselves into an all-consuming commitment which, because it is all-consuming, demands immediate return and when reality proves recalcitrant despair quickly sets in. As one of our interviewees put it, 'there's definitely a danger of tying your whole sense of worth and purpose to this challenge that is so much bigger than you and is never ending'. And this binary is often linked to another one which is 'now or never'. In climate change this is manifest in the belief that 'we must all act now or it will be too late'[6] a belief which can all too quickly slip into the perception that it is already 'too late', that processes have already been unleashed which are irreversibly leading us to catastrophe. This is the danger with the language of 'tipping points', it can tip the individual over into overwhelming despair so quickly.[7] Catastrophism has been a recurring motif in political movements for millennia[8] and it is no stranger to the politics of climate change.

One obvious and much parodied consequence of either/or thinking is the splitting and factionalism which often bedevils political groups and movements. As the Monty Python crew knew, it goes way back to the Old Testament. It reads, 'you're either with us or you're against us'.[9] This infects both reactionary and emancipatory political movements, and identity politics has been particularly susceptible to it, where the pressure to 'take sides' can be very powerful.[10] A contrasting attitude is one which insists that 'if you're not obviously against us then you are potentially with us'. It respects different views for being what they are, i.e. different but not opposite. It is not just tolerant of difference but embraces it, seeing in it the opportunity to learn from the different experience and perspective, thereby deepening one's own. It appreciates how an angler can also be a conservationist, and how a conservationist can also be a direct activist. It recognises how a Conservative or Republican advocate of small government can also be deeply worried about environmental destruction and an ardent advocate of frugal living.[11]

George Marshall has urged us to beware of what he calls 'enemy narratives' around climate change[12] and, given our anxious propensity to go creating monsters where none exist, this is a salutary reminder. He may be right, to talk of enemies does quickly take us into a paranoid place, but, to paraphrase the old saying, just because I'm paranoid doesn't mean that I don't have enemies.[13] It has become common to think of fight/flight as a kind of group pathology.[14] But it would be dangerous to then assume that the very idea of 'fight' is somehow wrong. To the contrary, the group when it functions creatively may still have to fight at times. The real challenge is to be able to have the right fight with the right opponent conveying the right message at the right time. When Just Stop Oil activists blocked busy motorways, or Extinction Rebellion activists prevented mostly working class commuters from getting to work on part of London's Underground, how many of these boxes did they manage to tick? Not many in my estimation. Or compare the powerful symbolic significance of throwing the statue of Edward Colston into Bristol's docks with the limp and confusing symbolism of throwing soup at Vincent Van Gogh's picture of Sunflowers.

Another binary that the climate movement sometimes falls into relates to its theory of political consciousness and change. 'Disruption' has been an absolutely central tactic of XR and other groups when blocking motorways or disrupting sports events. But there is a tendency for many within these movements to connect the tactic with another binary, between those who are awake and those who are asleep. Activists are awake, the general public are asleep and need waking up via disruptive tactics. This is an all or nothing view of political consciousness, one rejected by previous generations of activists who, influenced by the thinking of Antonio Gramsci, saw political consciousness as a contradictory phenomenon – we are all to some extent asleep and we are all to some extent awake.

So these are some of the binary oppositions that proliferate within political movements for change. Each one carries with it the potential for self-destruction, particularly in periods when a movement faces real setbacks which make the underlying despair hard to contain. If the movement to combat climate change is to be successful then all of us must overcome the pull towards thinking in either/or and 'all or nothing' terms and hold the tension between hope and despair.

Sustainable activism

What emerged from our interviews with the first wave of climate activists was that a new generation was emerging that was developing a much more emotionally intelligent activist culture, a kind of sustainable activism. This has continued with the second wave of climate activism around Extinction Rebellion. Direct action places activists in a vulnerable situation. Rather than a macho denial of vulnerability, the new generation of activists seems much more prepared to acknowledge this vulnerability through systems of formal and informal support such as organised debriefings after 'actions' and trauma support networks. Many activists also seem to be able to take a proportionate response. Intense engagement when needed can be followed by the capacity to take a step back and give attention to self-care.[15]

Some within this new generation are already talking about a sustainable activism, one that is in it for the long term. As one of our interviewees put it,

> The struggle will always be there for justice and for those kinds of things and I don't think we should...there's no utopic end point is what I mean.

It will always be evolving and changing ...there will always be another struggle somewhere...

A sustainable activism has a 'pessimism of the intellect' which can avoid wishful thinking and can face reality as squarely as possible, but it also retains an 'optimism of the will', an inner conviction that things can be different. It is hopeful but this is not so much a hope that you 'have' but an embodied hope that simply 'is'. The hope exists, despite all the reasons for acknowledging the situation as being hopeless.

Whiteford Lighthouse, the Gower

By holding the tension between optimism and pessimism a sustainable activism is more able to bear the despair and has less need to resort to splitting as a way of engaging with reality. It can hold the contradictions so that they do not become either/or type polarities. It can work both in and against the system. Whilst it believes there can be no personal change without political change it is equally insistent that there can be no political change without personal change. It insists optimistically that those who are not against us must be with us and therefore carries a notion of 'us' which is inclusive and generous, one which offers the benefit of the doubt to the other.

Finally it holds that it is never too late. In the context of climate change it is able to face the truth that some irreversible processes of change are already occurring, that even a two and a half degrees limit in the increase in global temperatures is unlikely to be achieved, that bad outcomes are inevitable and some are already happening. It nevertheless insists this makes our struggle all the more vital, vital to reduce the scale and significance of these future outcomes, to fight for the 'least worse' outcomes, and to ensure the world of our grandchildren and their children is as habitable as possible.

---

[1] Gould, D. (2009) *Moving Politics: Emotion and ACT UP's Fight Against AIDS*. Chicago: University of Chicago Press.

[2] An honourable mention must be given here to the work of James Jasper and Jeff Goodwin, two New York based sociologists, who have consistently called attention to the role of emotion in political movements. Goodwin, J., Jasper, J. and Polletta, F. (2001) *Passionate Politics: Emotions and Social Movements*. Chicago: University of Chicago Press.

[3] Hoggett, P. and Randall, R. (2018) Engaging with Climate Change: Comparing the Cultures of Science and Activism. *Environmental Values*, 27: 223-243.

[4] The concept of 'splitting' is key to Kleinian theory and practice. For an accessible introduction to Kleinian thinking see Segal, H. (1973) *Introduction to the Work of Melanie Klein*. London: The Hogarth Press.

[5] It follows that it is absolutely vital that the direct action wing (sometimes referred to as the 'radical flank') of the climate movement (exemplified by Just Stop Oil) and the 'moderate flank' of activists (involved in community energy, Transition initiatives and capturing local government) are seen as essential complements to each other. Each opens up space for the other to operate within.

[6] Parts of the climate change movement have adopted a favourite messaging device from the nuclear disarmament campaign, that is, that we are just a few minutes away from the midnight apocalypse. It didn't work back then so there's no reason why it should work now.

[7] Mann, M. (2021) *The New Climate War: The Fight to Take Back Our Planet*. London: Scribe. Chapter 8.

[8] Cohn, N. (1970) *The Pursuit of the Millennium*. London: Paladin.

[9] The famous sketch in the Life of Brian which is set in the Roman Amphitheatre depicts a heated argument between the People's Front of Judea and the Judean People's Front.
[10] Pilgrim, D. (2022) *Identity Politics: Where Did It All Go Wrong?*. Biecester, Ox: Phoenix.
[11] See, for example, the extracts of an interview with Jill, a Republican from Florida, reported by Renee Lertzman. Lertzman, R. (2019) New Methods for Investigating New Dangers. In Paul Hoggett (Ed.) *Climate Psychology: On Indifference to Disaster*. Basingstoke: Palgrave Macmillan. George Marshall also shares some illuminating experiences and conversations about the environment and climate change he had with Tea Party Republicans. Marshall, G. (2014) *Don't Even Think About It: Why Our Brains Are Wired to Ignore Climate Change*. London: Bloomsbury Academic.
[12] Marshall, Don't Even Think About It, see note 11 above.
[13] Andreas Malm offers a provocative counterpoint to the predominant paradigm of non-violent direct action which has guided the climate movement through Extinction Rebellion etc. He asks us to honestly question whether NVDA, even on a mass scale, is sufficient to bring about the transformation needed. He argues that the dire situation requires that the movement shifts from protest to resistance. 'Protest is when I say I don't like this. Resistance is when I put an end to what I don't like'. Malm, A. (2021) *How to Blow Up a Pipeline*. London: Verso. p.70.
[14] For Bion group life was characterised by three nonrational modes of functioning, one of these he termed 'fight/flight'. Bion, W. (1961) *Experiences in Groups*. London: Tavistock Publications.
[15] Charlie Wood, an Australian climate activist, offers a very engaging, insightful and personal account of burnout and the lessons to be drawn from it. Available at: https://wagingnonviolence.org/2022/08/no-place-for-burnout-in-a-burning-world/ (accessed 6 September 2023).

CHAPTER TWENTY-THREE

WHAT IS MORE AND WHAT IS LESS?

Ecological austerity

Until recently the idea of a 'no growth' or 'degrowth' economy was regarded as a policy choice championed by deep greens and ridiculed by those who liked to pose as realists. If we are to avoid dangerous climate change, it was argued, then we (in the developed western societies) must abandon our carbon intensive lifestyles and adopt simpler ways of life. Now it is beginning to dawn on me that this is no longer a choice that we could adopt; it is likely going to be forced upon us. To use the language of 'austerity' that has crept into our vocabulary since the 2008 global economic crash, exhausted nature is about to impose its own brand of austerity upon us – *ecological austerity*.

Research on human resilience distinguishes between sources of stress, such as a bereavement, which are episodic and those, such as poverty, which are chronic and enduring. Covid constituted an episodic threat. Climate change is a chronic, cumulative, enduring and deepening threat. It is no longer simply an emergency; it has become known as 'the long emergency'. It is here to stay. Perhaps to live during this long emergency will be like living with a chronic or degenerative illness?

The symptoms of muscular dystrophy (MD) commonly first appear in childhood; there is no 'cure' just a prospect of increasing disability. It can be fatal if it affects the heart. However, there are various forms of intervention which can slow, even occasionally halt, the process of degeneration and of course there are support groups. I have a friend whose MD was diagnosed over 35 years ago. He is now a head teacher at a state secondary school, he's married and has a daughter. He's had pain all this time, it's severely affected the straightness of his spine and shoulders. He now walks with great difficulty, hasn't been able to run or climb for decades. Has he had a good life? He would undoubtedly say yes.

When I was at university there was a young philosopher who I got to know called Havi Carel. She had been unwell for some time, and was eventually diagnosed in 2008 with a degenerative respiratory disease called LAM. When Havi was diagnosed she was given 10 years to live. She's still alive and now Professor of Philosophy at Bristol.

In her short, moving and accessible book,[1] Havi poses the question. It is a question for all of us: can we (both individually and societally) be ill and happy? Havi wrote that at first 'the inevitability of decline was the only principle governing my life'[2] but interestingly rather than raging against this dying of the light she says 'I surrendered, accepted, began to make room for my inability'.[3] Steeped in Ancient Greek philosophy, Havi noted how her condition meant 'I had to rethink my idea of a good life'.[4] She reflected that 'the fragility, but also the preciousness, of the present became a fundamental building-block of my experiences'[5] enabling her to see 'each day and each moment as an unexpected gift'.[6]

Returning to the question Havi Carel poses – can we be ill and happy? – the answer is most definitely 'yes'. But to crib from a famous phrase 'this is happiness Jim, but not as we Moderns mostly know it'. The metaphor of a chronic and possibly degenerative illness could be an apt metaphor for the long descent of climate change that we have begun to embark upon if it were not for one difference. The metaphor construes illness as an act of fate. But climate change is not an accident, it is something we are collectively responsible for (and some much more than others). So a more apt metaphor would be lung cancer or an equivalent disease where the misfortune is something we have brought upon ourselves.[7]

What isn't counted doesn't count

The growthism of the global North currently threatens to irreversibly damage the planet that we live on. Growth is measured by GDP, Gross Domestic Product. I visit the UK government website to ask the question 'what is GDP'? And it tells me, it says 'GDP matters because it shows how healthy the economy is'. Mainstream political parties compete to present themselves as the most likely to increase a nation's GDP. And a universal but palpably false consensus exists that the only way of improving the living standards of the less well off in society is by increasing GDP. The idea that wealth might actually be redistributed is seen as a belief nowadays only held by unrealistic dreamers.

In reality GDP includes everything that passes through the market, whatever the nature of the particular commodity and regardless of its wastefulness, destructiveness, irrationality, and the inequality, exploitation, and expropriation embedded in it. To give an example, consider that just prior to the COVID-19 pandemic one significant industry sector of the UK economy contributed £14.5 billion to Britain's GDP and almost £3.2 billion per annum in taxes to the Treasury. And because it is part of the service economy it is not involved in an extractive relation to the natural environment. Sounds good, doesn't it? In today's newspeak this sounds like a sector we should want to see 'grow the future'. The only trouble is that Public Health England estimated that the sector was associated with 409 suicides in 2021. The sector in question is the UK Gambling Industry – a NHS survey indicated there were 245,000 problem gamblers (i.e. addicts) in 2018.

GDP is based on the exchange value of commodities not on their use value, hence the recent comment by Adair Turner, the Chair of the Financial Services Authority, that much of the banking sector was engaged in 'socially useless' forms of production.[8] The use value or social value of an activity or product depends on its intrinsic worth rather than its price. As we saw in chapter 2 to give an estimated price to something like the Great Barrier Reef is to anticipate its superfluousness; because its intrinsic value is immeasurable it is written out of such calculations and the so-called price that Deloitte's came up with was in reality the reef's write-off price at that time.

But what doesn't count towards GDP? Well, what are called 'transfer payments' (e.g. state pensions, unemployment benefit) for a start. But clearly these are public goods, and if a society had introduced a Universal Basic Income then seemingly that wouldn't count either. More crucially goods and services which are not monetised don't count. So if you look after your children and elderly relatives at home this doesn't count (in other words, much of the labour traditionally labelled "women's work" doesn't count) but if you put them in a private nursery or residential home then it does! I say it does count, but it doesn't count very much. You only have to consider the scandalously low wages earned by those in the care sector compared to many other sectors.

GDP only counts what can be measured and if it can't be counted then it doesn't count. The economist Jason Hickel usefully reminds us that Robert Kennedy put it in 1968 like this: 'GDP measures neither our wit nor our courage, neither our wisdom nor our learning, neither our compassion nor our devotion to our country....it measures everything, in short, except that which makes life worthwhile'.[9] Hickel argues further that '(a)fter a certain point, which high-income nations have long surpassed, more GDP adds little if anything to human flourishing'.[10]

In this book I hope I have made a sustained case for the way in which neoliberal society has a corrosive impact upon the adult self and social relations. Although this is not an area of expertise for me it seems the outlook may be even bleaker for our children. Back in 2008, widely reported research by Unicef indicated that, for example, despite belonging to the fifth largest economy in the world, British children were among the most unhappy in OECD countries.[11] This trend seems to be continuing; there are increasing signs that in strongly neoliberal societies such as USA and UK, economic growth relates negatively to vital aspects of wellbeing. Thus research conducted by Essex University in conjunction with the Children's Society has found that addiction, self harm, suicide and other indicators of mental ill health in younger people were increasing significantly even in the decade before the onset of COVID-19.[12] Social media appears to be a significant factor. A recent survey of over 1,000 children by stem4, a mental health charity, found disturbing levels of body image issues and associated eating disorders.[13]

Rethinking wellbeing: Human capabilities

Imagine a nation which has a thriving civil society (a vast array of groups, clubs and associations), dense and extensive household, kinship and neighbourly networks, and a close and symbiotic relationship to surrounding ecosystems. None of this would count in terms of GDP. For example, on any given night there are literally thousands of music gigs occurring throughout the UK, mostly in non-commercial venues. This crucial dimension of British cultural life which stretches back over 50 years only 'counts' when it produces exchangeable commodities – likewise for organised sport, hobbies, the arts, health and care, and so on.

So if beyond a certain point higher GDP begins to have a negative relation to human wellbeing how should the wealth of a nation be measured?[14] This has been a major area of rethinking at national and international level. For example, the Sarkozy Report led by Joseph Stiglitz[15] noted that GDP mainly measures market production rather than government or household provision of goods and services. This links to longstanding feminist critiques of the concepts of work and production for the way in which they ignore domestic labour.

In thinking of an alternative model of human wellbeing many greens have been drawn to what is called 'the capabilities approach'.[16] This has adapted the Greek philosopher Aristotle's concept of eudaemonia or 'the good life'. A capability refers to an opportunity to do or be something, for example, to be able to go out in the evening and mix with others. Now this might seem straightforward enough but as the north American philosopher Martha Nussbaum noted, it is in fact a freedom denied to hundreds of millions of women in Asia and the Middle east.[17] Clearly then there is a close relationship between a capability and a freedom to do or be something. Nussbaum suggested[18] that there are ten universal capabilities, some of which relate to 'freedom from' and some to 'freedom to'. They are:

> *Life:* Being able to live a satisfying life into old age.
>
> *Bodily Health*: Being free from ill health which impacts on quality of life.

PARADISE LOST? The Climate Crisis and the Human Condition 195

> *Bodily Integrity*: Being able to have control over one's own body and its functioning.
>
> *Control Over One's Environment:* Being able to have freedom of expression, ownership, employment.
>
> *Senses, Imagination & Thought:* Being able to enjoy/participate in cultural experiences such as music or storytelling.
>
> *Emotion:* Being able to enjoy the full rich range of positive and negative emotions.
>
> *Practical Reason:* Being able to think critically about the world and one's place in it.
>
> *Affiliation:* Being able to associate with others.
>
> *Other Species:* Being able to engage with other species and plants.
>
> *Play:* Being able to enjoy playful relations (formal and informal) with others.

The emphasis upon human freedoms offers a progressive perspective for many groups whose lives are bound by social or economic restrictions. However there are also major problems with this approach. Its emphasis upon freedom from limits and constraints is very congruent with liberal and neoliberal values concerning the sovereign individual.[19] What about the other side of freedom? What about obligation and responsibility, both to others and to nature? Why isn't there a list of capabilities which relate to interdependency, care, compassion, acceptance, equanimity, receptivity, and so on.

I wonder if this raises another problem. To talk about freedoms and responsibilities still places the individual human at the centre of things. But isn't this why we are in the ecological and social crises we are in? Traditional societies do not place the individual at the centre of things, they place the extended family, kinship group, tribe or clan at the centre. Many societies go much further, dissolving the boundary between the human and non-human so that the status of 'person' is extended to animals, plants and rivers.[20] Sounds a bit mystical? Well, modern capitalism already grants the status of personhood to things which are complete abstractions, i.e. to companies and firms, and what could be madder than that?

To return to the capabilities approach, we must reflect upon the deeply individualised way we think about our relations to others let alone our relations to nature. As we saw in our discussion of love and kindness, by thinking of such things in terms of the active, autonomous individual and their choices we obscure much of what is vital about them. The modern individual self is always placed (and places itself) at the centre of things. It is time to stop.

Human capacities

Like the idea of vitality the concept of capacity is ubiquitous in psychoanalytic literature but it has hardly ever been subject to systematic exploration. My former colleague Robert French[21] noted that even Winnicott, who put forward important hypotheses about what he called 'the capacity for concern' and 'the capacity to be alone', never paused to reflect on what he had in mind when he used the word 'capacity'.

Robert noted that the etymology of the word was derived from the Latin 'able to take in' or 'able to hold much'. He recognised the many ways in which the concepts of capacity and capability seemed to overlap but that what was important about the former was the way in which it constituted a critique of, and contrast to, the dominant managerialist and modernist concept of human learning and development. This construed learning in terms of 'a mastery of a succession of competences or skills that are predefined... expressed, for example, in the omnipresent phrase "by the end of this module you will be able to..."'.[22] Robert noted the links of this model of learning and growth to the culture of performativity that I explored in Chapters 5 and 6, one which seeks to measure, compare and standardise everything; he thought of it as the McDonaldization of learning.

Robert did not want to dismiss the importance of skill development and noted that, for example, the psychotherapist needs to learn skills such as listening and how to offer an interpretation. But what was also crucial were a set of capacities that could not be learnt in the way in which a skill could be learnt – for example, the capacity to stay with uncertainty[23] or to be able to move from a close state of identification with someone to disidentifying with them (as any parent must learn to do). These capacities are multi-dimensional involving thinking, feeling, being and doing. Perhaps most importantly *capacities are relational*. They cannot be 'taught', but they can be 'developed' through experience (and for trainee therapists this includes the experience of being in therapy oneself).

So how does a capacity relate to a capability? For Nussbaum a capability refers to a freedom to be or do something, my suggestion is that a capacity refers to our potential to then make use of that freedom. Consider 'affiliation', that is, the ability to associate with others. As Nussbaum notes, in many cultures women's freedom to affiliate, to mix and associate with others, is severely restricted by law. However in many liberal western societies although this freedom to affiliate formally exists, levels of violence against women are such that many women now feel their freedom to affiliate, particularly at night, is very restricted. Moreover these societies have become so hyper-individualised that many people, male and female, have lost their

PARADISE LOST? The Climate Crisis and the Human Condition   197

capacity to engage in sociable ways with others. These days many people who come seeking therapeutic help come precisely because they have lost the basic trust to be with others in a non-anxious, unselfconscious way. They approach social encounters on high alert, constantly vigilant and on the lookout for slights, hurts, rejections or provocations. In other words, their capacity for affiliation has become severely diminished and this has been strongly influenced by the wider culture, and the most startling evidence for this is the place where most attempts to affiliate take place nowadays – online.

Consider another of Nussbaum's capabilities – emotion – in this light. I have examined the way in which contemporary neoliberal society casts a shadow over the emotional lives of its citizens, fostering negative emotions such as shame, resentment and envy and fracturing our internal worlds by encouraging the splitting of thought from feeling. This makes it so much harder for us to engage in emotionally complex ways with others. For example, to feel compassion towards those we dislike or disagree with, to tolerate uncertainty without being overwhelmed by anxiety, or to find as much pleasure in the company of those who are different to us as we do in the company of those who are similar. It follows that a capacity refers to a power or potential which can be realised under some social conditions more than others. Such social conditions are equivalent to a facilitating environment.[24] A good society brings out the best in us, a bad society does the opposite.

Containing contradictions

Let's return to Robert French's reflections on the original meaning of the word 'capacity', that is, to be able to hold much. One of the most crucial psychological capacities that we have is to be able to hold contradictions and, as we saw in the previous chapter, this has important consequences for how we think about nature and society and engage with the climate emergency.

I do not believe we can resolve the dichotomies that characterise the modern human condition – human/creature, civilization/nature - by simply dissolving the boundaries and differences which are posed by them.[25] To be able to hold a contradiction such as human/creature without creating a warring polarity seems to me to be the task both of each single individual and of humanity.[26]

What does this mean in practice? Take the idea of progress, a concept integral to the modernist imagination. I have highlighted the uniquely destructive potential of modern civilization and the way in which humanity has now become a force of nature. Unsurprisingly many within the environmental movement have little if anything positive to say about progress – the advances of science, technology, urbanisation, etc are all viewed with profound scepticism. Concepts such as 'green growth' are seen as contradictions in terms, and technological solutions to the climate change problem (the so called negative emissions technologies such as high atmosphere aerosols which would reflect the sun's rays back into outer space) are (quite correctly) seen as potentially more dangerous than the problem they are designed to address. In contrast, the premodern is positively valued, if not idealised – indigenous wisdom, natural food/health/childbirth, small-scale rural self-sufficiency. It comes as no surprise then to see some influential representatives of what might be called the deep green milieux, such as Charles Eisenstein and Paul Kingsnorth, adding fuel to vaccine scepticism because of their rejection of (at least some aspects of) modern medical science.

But nature good/civilization bad is not the answer. Many indigenous practices such as female genital mutilation we rightly regard from our civilized perspective as barbaric. Over the centuries what Norbert Elias called 'the civilizing process'[27] saw massive reductions in the violence of everyday practices including child rearing, sport, entertainment, the treatment of animals, criminal punishment, etc. For myself, I love much of what modern civilization has given us. It has given my daughter freedoms that were not available to her grandmothers. It has enabled my brother-in-law to live free from persecution despite his sexuality. It was able to diagnose my cancer back in 2006 and then successfully treat it. It has given me my Roland FA08 keyboard and LogicPro music production software. It has given me my

PARADISE LOST? The Climate Crisis and the Human Condition

electric car and the solar panels on my roof, an app for identifying birds via their song and one for identifying plants via their photographic image. It has also given me and us all the things I mentioned at the start of this book – mass species extinctions, the disappearance of indigenous cultures, racism and genocide, and so on.

This is the contradiction. To say that there is a bit of good and a bit of bad in civilization is to opt for a false resolution for the good is very, very good and the bad is so extremely bad. We cannot 'resolve' a genuine contradiction but we can develop our capacity to hold it and the tensions that this holding generates. Things don't add up and it is precisely in staying with this not-adding-upness that the deepening of human moral, spiritual, aesthetic and practical capacities lies. Vitality within individuals and groups arises precisely because things don't fit, identities can't be integrated, viewpoints differ, dilemmas can't be resolved, predicaments can't be solved and conflicts smoulder on.

Robin Wall Kimmerer draws both upon her indigenous Potawatomi ancestry and her scientific expertise when towards the end of *The Democracy of Species* she says:

> I dream of a world guided by a lens of stories rooted in the revelations of science and framed with an indigenous worldview – stories in which matter and spirit are both given voice.[28]

Fantastic. This is what I mean by the containment of contradiction.

---

[1] Carel, H. (2008) *Illness: The Cry of the Flesh.* London: Acumen.
[2] Ibid, p.63
[3] Ibid, p.64
[4] Ibid, p.61
[5] Ibid, p.124
[6] Ibid, p.133
[7] I am very grateful to fellow CPA member Breda Kingston for highlighting this difference.
[8] Treanor, J. (2009) FSA boss Lord Turner attacks CBI chief over 'socially useless' City behaviour. The Guardian, 23rd November ,2009. Available at https://www.theguardian.com/business/2009/nov/23/lord-turner-cbi-fsa-city (accessed September 2023).
[9] Hickel, J. (2020) *More is Less: How Degrowth Will Save the World.* London: Heinemann. p.200
[10] Ibid, p. 174.
[11] UNICEF (2007) Child Poverty in Perspective: An Overview of Child Wellbeing in Rich Countries; https://www.theguardian.com/society/shortcuts/2012/jun/27/why-british-children-so-unhappy (accessed 6th September 2023)

[12] Campbell, D. (2021) Number of UK children unhappy with their lives rises – Report. The Guardian, 26 August 2021. Available at: https://www.theguardian.com/society/2021/aug/26/number-of-uk-children-unhappy-with-their-lives-rises-report (accessed 6 September 2023).
[13] Hill, A. (2023) Three in four children dislike how they look, The Guardian, 2 January 2023. Available at: https://www.pressreader.com/uk/the-guardian/20230102/282303914219051 (accessed 6th September 2023)
[14] There have been several attempts to identify this from a more ecologically sensitive perspective. It has, for example, been a continuing thread in the work of the New Economics Foundation.
[15] Stiglitz, J., Sen, A. and Fitoussi, J-P. (2009) Report by the Commission on the Measurement of Economic Performance and Social Progress. Available at: https://web.archive.org/web/20150720212135/http://www.stiglitz-sen-fitoussi.fr/en/index.htm (accessed 6 September 2023)
[16] Sen, A. (2005) Human Rights and Capabilities. *Journal of Human Development*, 6(2): 151-166.
[17] Nussbaum, M. (2005) Women's Bodies: Violence, Security, Capabilities. *Journal of Human Development*, 6(2): 167-183.
[18] Nussbaum, M. (2000) *Women & Human Development: The Capabilities Approach*. Cambridge: Cambridge University Press.
[19] See James, P. (2018) Creating capacities for human flourishing: An alternative approach to human development. In P. Spinozzi & M. Mazzanti (Eds.) *Cultures of Sustainability and Wellbeing: Theories, Histories, Policies*. London: Routledge.
[20] Robin Wall Kimmerer offers a subtly illuminating account of the contemporary implications of an animistic view of the world, arguing convincingly that 'the animacy of the world is something that we already know' as toddlers but have to 'grow up' and out of. Wall Kimmerer, R. (2021) *The Democracy of Species*. London: Penguin. pp.15-21. Amitav Ghosh's book *The Nutmeg's Curse* is in a way an extended paean to this animistic way of being in the world. Ghosh, A. (2021) *The Nutmeg's Curse: Parables for a Planet in Crisis*. London: John Murray. See also Hickel, note 10 above, Chapter 6.
[21] French, R. (1999) The importance of capacities in psychoanalysis and the language of human development. *International Journal of Psychoanalysis*, 80: 1215-1226.
[22] Ibid, p.1220.
[23] Interestingly enough this is known in the psychoanalytic literature as 'negative capability'.
[24] The title for a key collection of papers of D.W. Winnicott. Winnicott, D.W. (1965) *The Maturational Process and the Facilitating Environment*. London: Hogarth Press.
[25] Andreas Malm refers to this as 'the dissolutionist crusade'. He sees figures such as Donna Haraway and Bruno Latour at the front of this crusade, the charismatic flag bearers in the fight to dissolve all polarities, oppositions and binaries. Malm, A. (2020) *The Progress of this Storm: Nature and Society in a Warming World*. London: Verso. pp. 186-190.
[26] As I mentioned in Chapter 2, compared to Haraway and Nietzsche only Ernest Becker seemed able to stay with and hold this human/creature contradiction.
[27] Elias, N. (1994) *The Civilizing Process*. Oxford: Blackwell.
[28] Kimmerer, see note 20 above, p.85.

CHAPTER TWENTY-FOUR

WELLBEING

Relational beings

Human beings are relational beings, that is, they find fulfilment by virtue of their relations with other human beings and in their engagements with nature and the material world.[1] It is upon the quality of these relations that human capacities depend. This idea, if pursued consistently, has radical implications both for our vision of a "good society" and for our conception of politics. It means, for example, that we are absolutely dependent upon nature, including our own creaturely nature, for our wellbeing. We are also absolutely dependent upon other human beings and, in an identical fashion, others are dependent upon us. Such simple ideas but how hard we rebel against them! The individualism of contemporary society pitches us against the very ideas of humility and respect that interdependency implies.

Our relational capacities are manifest in the quality of our affiliations, affections and attachments. They are reflected in our capacity to think beyond self and thus in our capacity for care and concern. They are also manifest in our capacity to manage our own internal conflicts and difficult feelings without projecting them onto others. But interdependency does not synonymous with cooperation and harmony. Our relational capacities are developed through the experience of difference, hostility and conflict as much as through solidarity and empathy, and this applies as much to our relations with nature as it does to our relations with humans. My father used to tell a story. His father, my grandfather, was a trawlerman who worked on the North Sea. Just before he died of cancer at the age of 84 my father visited him in hospital. Trying to comfort him my father said 'tell me about the good times at sea dad'. 'There were no good times at sea son' came the reply. Nature can be the source of wonder and magic; it can also be a source of huge adversity and suffering that we have to constantly struggle with.

Building on the thinking of Robert French I believe that our relational capacities manifest themselves in three areas of life that are essential to our wellbeing – the aesthetic and spiritual, the practical and intellectual, and the moral/ethical. These constitute much of what it means to be human, though clearly the form in which they are realised will be historically and culturally specific.

Aesthetic/Spiritual

Moral/Ethical   *Practical/Intellectual*

One of the tragedies of modern civilization is that it developed these capacities unevenly – whilst enormous importance was given to the practical and intellectual our moral, aesthetic and spiritual capacities remained undeveloped.[2] As I have argued earlier, this has encouraged a rationality which knows the price of everything and the value of nothing, that is, a purely instrumental way of thinking about humanity and nature solely as a means to an end.

Our moral and ethical capacities appear to be severely undernourished. I have been very struck by Dan Gretton's exploration of the idea of the 'desk killer', of how easily good parents, partners, lovers and friends can become monsters in the public sphere and, conversely how public saints can be private monsters. I have never properly understood the distinction made between morality and ethics but it seems to me that whereas the former refers primarily to our private lives, the latter refers to the public sphere. I believe that psychoanalysis and the psychotherapies reflect this underdeveloped and fractured landscape given their preoccupation with peoples' private lives and disinterest in their public selves. Is our capacity for care and concern something only of relevance to our private lives? I think not. So why do the psychotherapies restrict their focus solely to this sphere?

Regarding our aesthetic capacities, Winnicott developed a relational account of the child's capacity for play, a capacity which he thought lay at the heart of the development of culture. Play contributes both to our aesthetic powers (in music, the arts, sport and other areas where we put our imagination to work) and to our conviviality; it is no coincidence that we speak of playing an instrument or playing football. Through aesthetic activities we seek to give representation to the sublime, to that realm of experience which is beyond words, and it is at this point that the aesthetic and the spiritual often overlap. This was the function of the shaman and of the many ancient and traditional rituals and practices that have struggled to survive under the onslaught of modernity.[3]

In speaking of the realisation of human powers and capacities I have in mind a potential which may or may not be realised. Winnicott used the phrase 'the facilitating environment'[4] to refer to the conditions that enable these capacities to flourish. I see this having physical, familial, social and cultural dimensions. How rich and diverse is the physical environment that we have access to? How nurturing and stimulating is the family environment in all its diverse forms? How safe, friendly and well-resourced are the neighbourhoods in which spend our days? To what extent do our cultural values and mores support and sustain the flourishing of moral, ethical, spiritual and aesthetic capacities as well as those necessary for economic progress?

An Ecowelfare Society

A relational perspective can help us understand more clearly how an ecowelfarist society would differ from, and offer more than, either consumerist or traditional welfarist models. How do these three visions of the "good society" differ?

Consumerism: Wellbeing lies in the quantity and variety of material goods and services purchasable as commodities by consumers.

Welfarism: Wellbeing lies in the quantity (if not variety) of public goods and services received as a right by citizens.

Ecowelfarism: Freedom from material need is a necessary but not sufficient condition for wellbeing, for the essence of a good society resides in the quality of the relations between people and between people and nature.

Consumerist and welfarist ideologies are primarily concerned with issues of quantity, and the concept of 'standard of living' is central to this. Economic growth, in turn, is seen as essential for delivering peoples' standard of living.

But to drive growth a constantly expanding horizon of commodities must be produced, new forms of longing induced and new realms of lack and insufficiency generated. In contrast an ecowelfare society would focus on the quality of social and environmental relations. It would privilege questions of quality over quantity once freedom from basic need had been secured.[5] Imagine schools where the quality of social relations between children, and between children and teachers, was put at the heart of the educational project and all the implications of that for developing capacities for care and respect, for firing the imagination and driving curiosity. Imagine living in a neighbourhood which enabled people nearing the end of their lives to flourish rather than languish in isolation and a home care service where the quality of the relationship between the care provider and the user was central? What would such a neighbourhood look like, what are the physical design and planning implications, etc.? The connection to current arguments about 'the 15-minute city' are clear. How interesting that such ideas are construed by neo-illiberals as an attack upon fundamental liberties, and specifically the liberty of the motorist, that ultimate symbol of the isolated modern self.

Eco-democracy

We must recognise the presence of an authoritarian inflection within some aspects of green thought, one which has emerged as the distant threat of climate change has become the real and present danger of the climate emergency. It is one that assumes that people will only change if they are forced to. It relies on apocalyptic fear as a motivator and the green future it envisages is often perceived as austere and unattractive – a necessary medicine rather than something beneficial or fulfilling. To those living in the West it prescribes 'less' but says little about an alternative vision of 'more' beyond a greater connection to nature. Some describe this as 'green frugality'.[6] It is probably latently anti-democratic. It says that, given the imperatives we face, strong government may at times be needed to impose rules on recalcitrant businesses and citizens[7] (setting limits, for example, to a range of human behaviours such as choice of family size). This kind of society may therefore be less respectful of human autonomy than liberal democracy. Perhaps this is the price we have to pay if we are to avoid catastrophe in the future?

Climate despair can also lead to misanthropy, that is, to a barely disguised contempt for the human race. In Section 2 of this book I noted that as the climate emergency worsens so nationalist 'solutions' will become increasingly difficult to resist, and as we have seen this perspective is already being adopted by influential figures such as Paul Kingsnorth. But even within more mainstream elements of the green movement one can discern visions of the future which are unrealistic and potentially oppressive. Michael Saward noted, 'most visions of green democracy are...variants...on a model of direct democracy in small, often rural, face-to-face communities, characterised by labour intensive production, self-reliance if not self-sufficiency, a related minimisation of trade and travel between communities, and decision making by face-to-face assemblies'.[8] But this is dangerously close to a vision of a network of homogenous communities in a society stripped of complexity, where there is no need for the state and no need for welfare institutions. One often gets the sense that care, education, etc. would be informalised in a green utopia, and we know what that probably means in terms of gender relations. There are striking parallels between the notion of 'return to community' and the old Marxist idea that in a post-revolutionary epoch there would be a withering away of class society and therefore no need for government or judiciary. The idea of people living in unity and harmony (i.e. a return to paradise) is a recurring theme in radical movements, as if a desirable future would be one without difference, conflict or antagonism, that is, one where there would be no need for politics.

In contrast to the dystopia of green frugality it is vital that we envision the "more" that a green society can offer beyond what a consumerist or welfarist society can provide. I think that the idea of an Ecowelfare society, considered in its expansive and positive sense as that which provides the conditions for human flourishing, is central to this vision. I also think that such a society would generate forms of strong democracy built around a politics of interdependency that is unrecognizable to prevailing models of liberal democracy. This would not assume the rational, self-seeking, instrumental actor of conventional economics nor on the harmonious co-operator of some green utopias. Instead it would propose a fundamentally relational view of the human being in which both solidarity *and* conflict are viewed as inevitable and vital dimensions of a facilitating environment in which human capacities can flourish.

In modern pluralist societies, differences of culture, faith, values and lifestyle proliferate. These differences place governments, political parties, associations and social movements at the intersection of conflicting needs and alternative definitions of the common good.[9] In pluralist societies value conflicts are inherent and irresolvable. This means that democratic deliberation, whether in local face-to-face communities, in cities, regions, nations or at the level of the global community, will be an ongoing, never finishing process of argumentation. But this doesn't have to be the commodified, diminished and sometimes perverse notion of party politics that passes for democracy at the moment. Deliberative innovations, particularly at the neighbourhood level[10], have spread like wildfire over the last three decades[11] as a consequence of the influence of new social movements and indigenous peoples' struggles. These innovations are often relationally informed and have imported methodologies from group dynamics and conflict transformation. This new way of doing democracy could be thought of as a 'democracy of the emotions'.[12] Innovations in deliberative democracy are now being demanded by the climate movement, indeed the third of Extinction Rebellion's basic demands was for a Citizens Assembly on climate and social justice. This found an echo in the decision by the Scottish Government to implement a citizen's assembly on climate change which was independent of government but reported to it.[13]

---

[1] Of course, this begs the question: what do we mean by such concepts as nature, human, etc? I include within the human all those human artifacts that are the outcome of our actions through history. So the clock on my desk, the Apple Mac in front of me,

the 'cloud' (presumably a mass of servers located somewhere in the Arizona desert) on which much of my computer memory is stored, these are all part of the human. Now an objection might be made that there is now nothing on earth left untouched by humanity, indeed that this is implied in the idea of the Anthropocene. But nature doesn't disappear simply because it is no longer 'pure'. Indeed, as I sit here with my left hand over my mouth looking at what I have just written bacteria swarm across my fingers and blood pulses through my veins. Even such a sophisticated human assemblage as myself remains a creature of nature! So, everywhere nature invades the human, just as everywhere the human invades nature.

[2] The great social philosophers Theodore Adorno and Max Horkheimer perceived in capitalism the ultimate realisation of an instrumental rationality which enabled our practical/intellectual powers to flourish while our ethical, spiritual and expressive capacities remained undernourished. Adorno, T. and Horkheimer, M. (1979 [1944]) *The Dialectic of Enlightenment*. London: Verso.

[3] See for example Eliade, M. (1964) *Shamanism: Archaic Techniques of Ecstacy*. London: Routledge & Kegan; Tucker, M. (1992) *Dreaming with Open Eyes: The Shamanic Spirit in Twentieth Century Art and Culture*. London: Harper Collins.

[4] Winnicott, D.W. (1971) *Playing and Reality*. London & New York: Routledge.

[5] And some combination of public provision for basic needs combined with a guaranteed minimum income seems to be the method of achieving this.

[6] Togerson, D. (2001) Rethinking politics for a green economy: A political approach to radical reform. *Social Policy and Administration*, 35(5): 472-489.

[7] Whilst I really appreciate much of what the philosopher John Foster writes I have found that his recent 'tragically-informed' call for a political movement that is prepared to go beyond existing democratic norms to save the world from ecocide is a step too far for me to contemplate at the moment. Foster, J. (2022) *Realism and the Climate Crisis: Hope for Life*. Bristol: Bristol University Press.

[8] Saward, M. (1993) Green democracy? In A. Dobson and P. Lucardie, (Eds.) *The Politics of Nature*. London: Routledge. p 71.

[9] Timothy Snyder puts it like this: 'The point of politics is to keep multiple and irreducible goods in play, rather than yielding to some dream, Nazi or otherwise, of totality.' *Black Earth* is a study of the Holocaust in Eastern Europe. It concludes with a warning that climate change is generating the conditions in which totalitarian dreams begin to flourish once again. Snyder, T. (2015) *Black Earth: The Holocaust as History and Warning*. London: Bodley Head.

[10] Innovations in neighbourhood democracy in the UK became associated with the response of radical local governments to Thatcherism in the 1980s. Burns, D. Hambleton, R. and Hoggett, p. (1994) *The Politic s of Decentralisation*. Basingstoke: Macmillan. Recently a new movement for neighbourhood democracy has sprung up in the UK. Visit their website at https://localis.org.uk/research/renewing-neighbourhood-democracy-creating-powerful-communities/ (accessed 12th October 2023).

[11] Paradoxically many of these initiatives have arisen partly because of the moribund and elite controlled nature of conventional democracy and as a consequence they are in constant danger of being recouped by these conventional forces. Hammond, M. (2020) Democratic innovations after the post-democratic turn: between activation and empowerment. *Critical Policy Studies*, 15(2): 174-191.

[12] Hoggett, P. and Thompson, S. (2002) Democracy of the Emotions. *Constellations*, 9 (1): 106-126.

[13] Scotland's Citizens Assembly on Climate Change met between November 2020 and March 2021. For the report on its process and impact see https://www.gov.scot/publications/scotlands-climate-assembly-research-report-process-impact-assembly-member-experience/. (accessed 6 September 2023). Nadine Andrews, who was heavily involved in this initiative, undertook an informative review of the emotional experience of the participation of the 106 members of the assembly. See Andrews, N. (2022) The Emotional Experience of Members of Scotland's Citizens' Assembly on Climate Change, *Frontiers in Climate*. Available at: https://www.frontiersin.org/articles/10.3389/fclim.2022.817166/ (accessed 24 September).

CHAPTER TWENTY-FIVE

ORDINARY ECSTASY

Just being

In our accelerating, time-compressed, hyper-individualised world it has become increasingly difficult for many of us to know how to just be. As anxious, vigilant neoliberal citizens we peer out from our psychic retreats, ready at any moment to withdraw in hurt and ruminate upon the injustices of life, goaded constantly by the fantasy of an ideal world which seems forever out of reach. In flight, fight or freeze for much of our lives we sadly seem unable to take in and engage with the ordinary extraordinariness of life. In contrast Havi Carel described how acceptance of her life-threatening illness had a regenerative effect on her emotional and personal life, enabling her to see 'each day and each moment as an unexpected gift':

> Ceasing to experience everyday moments as mundane, the present as tedious and the instant as insignificant is the way to achieve happiness now, immediately, in the present.[1]

The psychoanalyst D.W. Winnicott was acutely aware of the importance of being (or what today we would call 'being in the moment') and saw his approach as a contrast and possible compliment to traditional psychoanalysis with its focus upon desire and drive and therefore upon doing (and being done to).[2]

For Winnicott, the prototype of the experience of being is the development of the capacity to be alone during infancy, or as Winnicott puts it, to be alone in the presence of the other. This is an other who, although not necessarily always physically present, is experienced nevertheless as a reliable presence, one which instils confidence in the medium in which the child is held and therefore in the present and the future. Eventually, as an 'inside' and an 'outside' develops and the child becomes individuated,[3] this presence will be internalised, the child will have acquired the capacity to be alone without being lonely. For Winnicott this is the basis for the quiet sense of being which is so important to people's mental health. Not only this, he argues it is also essential to our capacity for play, something he contrasts to purposive activity. I watch my little grandson playing with two cars. He is talking to them, and the cars are talking to each other as they travel along the back of our sofa. He is lost in reverie and unaware of my presence. For

Winnicott, play is essentially improvisational, more than an activity, it is an experience which the child makes up as he goes along. He sees play as a form of agency, rooted in being, which manifests what he calls the 'female element' in human functioning.[4]

In a kind of manifesto statement, Winnicott once said, 'After being – doing and being done to. But first, being'.[5] He wondered whether there was such a thing as a form of ecstasy which was linked to being rather than to doing. Perhaps, he mused, the happy play of the child or the experience of an adult at a concert or the enjoyment of a friendship may be a kind of ecstasy.

I want to propose we refer to such experiences as forms of 'ordinary ecstasy'[6] and suggest that one of the most important sources of such experience arises from our engagement with nature. That wonderful poem 'An Ordinary Day' by Norman MacCaig captures it for me. Going for a walk he writes:

> ....Cormorants stood on a tidal rock
> With their wings spread out,
> Stopping no traffic. Various ducks
> Shilly-shallied here and there
> On the shilly-shallying water.
> An occasional gull yelped. Small flowers
> Were doing their level best
> To bring to their kerbs bees like
> Ariel charabancs. Long weeds in the clear
> Water did Eastern dances, unregarded
> By shoals of darning needles. A cow
> Started a moo but thought
> Better of it – And my feet took me home
> And my mind observed to me,
> Or I to it, how ordinary
> Extraordinary things are or
> How extraordinary ordinary
> Things are......[7]

To be able to live in and with time and with the other is to be able to mourn for a world which is constantly passing, to feel the ache of sadness as all that we see before us rushes by. I find myself often having to remind others that, far from being a negative emotion, sadness is an incredibly rich emotion with an intimate connection to beauty. When I visit my favourite oak tree or sneak a glance at the woman who has been my partner for over forty years or when I used to gaze in wonder at my baby grandson, I see something which touches

PARADISE LOST? The Climate Crisis and the Human Condition

my heart precisely because of its transience. But loss is only half of the picture for it is also true that to live in and with time is to feel joy for a world which is constantly in the process of becoming. There is no mortality without natality. This is the extraordinary nature of the ordinary, something which I believe most of us for some of the time are attuned to.

George Shaw, the Turner Prize nominee, grew up on the Tile Hill Estate, on the edge of Coventry. Sometimes called the Rembrandt of the Council Estate he paints the semi-detached houses, unloved open spaces, abandoned pubs and rows of off-street garages of the area where he grew up. As Sean O'Hagan put it, he 'records the mundane, the quotidian and the overlooked'.[8] His paintings have a sad, ominous and eerie quality, but they are also full of grief and beauty. Speaking of one of his paintings called *Scenes from the Passion: The Fall* one critic remarked that it featured 'a row of tatty lock-ups, unloved and partly disused…reflected in muddy puddles and overshadowed by winter trees against a yellow sky.'[9] Somehow, he brings a luminous quality to these pictures of the ordinary whilst at the same time providing a political portrait of decline and hopelessness.

George Shaw: *Scenes from The Passion: The Fall, 1999* Humbrol enamel on board 75 x 100 cm, Private Collection, London. Copyright The Artist / Courtesy Anthony Wilkinson Gallery, London

In praise of the ordinary

Dolton, a village in Devon, has a free monthly diary, 40 odd pages of adverts, reports and village information. Among the regular columns one is simply signed AndrewtheGardener. Andrew describes himself as a retired gentleman who loves gardening. Recently Andrew has adventured into beekeeping and his exploits have found plenty of coverage in his column. Here are some of his early reflections:

> The hive is in my shed connected to the outside world with clear tubes that make observation easy. Outside the shed there is chaos. There is a mass of bees all waiting to fly in or out. I made a little landing platform for them. The returning bees are so full of nectar they can hardly fly, some bees are good shots flying straight into their tubes, some land on the platform whilst others make no attempt at accuracy they just seem to stop flying and fall onto the ground when they get near the hive entrance, after a few seconds on the ground they fly up and have another go. Some return with different colours of pollen, green is particularly en vogue at the moment because the blackberries are flowering but they alight with white, (possibly elder), yellow, (maybe rose), orange perhaps buttercup, but brown, What could that be?
>
> Huge chunks of my days are devoured by these bees, it is addictive watching them flying in and out. I sit in a wheelbarrow resting its handles on an old car tyre, I am very comfortable and happy as a sandboy.

Just to the south of Newport, set on the Severn flood plain there's a hamlet called St. Brides. Just before the village there's a lane taking you to the Lighthouse Inn which is right next to the four-metre-high levee which stops the sea from flooding much of the land along the Severn Estuary.

Beyond the pub you'll find the Lighthouse Chalet Park, a magical little place with the levee on one side and the rhines and ditches of the flood plain on the other. Probably sixty or so chalets altogether and mostly much loved. Each with busy but well-maintained gardens tended by late middle-aged men with gnomes, hanging baskets and bird feeders everywhere. One or two have terrific topiary and where one of the roads ends by a drainage ditch a couple of residents have built elaborate bird boxes mounted on eight-foot-high white posts, which are much used by the smaller members of the avian species.

Being with others

I believe there is a very close connection between the capacity for just being, being in time, and the capacity for being with the other (whether human or nonhuman). At its root conviviality relates not to what we do with others but to our capacity to be with others – to derive pleasure from companionship, conversation and play with others. Anyone who looks at the organizations of civil society – choirs, clubs, etc. – only from an instrumental perspective, seeing them simply as a means to an end, misses out on half of what they are about. People derive tremendous fulfilment from just 'being with others' and the ostensible purpose of group activity often almost becomes secondary.

Back in the 1980s the very first piece of academic research I undertook involved a funded two-year study of the groups that people form through their passion for the arts, sport, leisure, hobbies and so on. In Leicester and Kingswood (an eastern suburb of Bristol) we studied groups of aquarists, onion growers, military modellers, cavers, hockey players, lapidarists… the variety was kaleidoscopic. In the conclusion to the little book which reported on our research we said: 'The clubs and associations described in this book are not merely settings in which people "do" things… [T]he groups are places to "be" as much as to "do", with their own histories, characters, dramas and meanings'.[10] As I write this, I am reminded of that marvellous BBC TV series *The Detectorists* which lovingly imagined the characters of a metal detecting society somewhere in deepest Suffolk.

I'm an avid skittles player and belong to a team called the Spiders who, it is said, began before the Second World War as a team of clerks in Bristol's aircraft industry (clerks in those days used spidery handwriting to record transactions in their ledgers, hence our team name). In my experience much of the richness of life, including the quality of friendship, comes from these everyday engagements. It is this primary social medium which constitutes the joy of ordinary life, the going-on-being-with-others in families, friendship groups, clubs and societies which is the basis of all human cultures.

The Ideal and the Ordinary

Far from having been lost in the distant past, or something we will reach in the future with one last effort, paradise is in reality in front of us but tragically, for many, hiding in plain view. It is present in the mundane, the everyday and the ordinary. To know this, and to live this, is the kernel of meaning in that rather overused phrase 'to live in the moment'.

[1] Carel, H. (2008) *Illness: The Cry of the Flesh.* London: Acumen, p.133.

[2] Winnicott explores the child's (and therefore the adult's) capacity to be alone, something he sees as foundational to psychological wellbeing, in his 1958 paper. He then develops this approach in a series of papers examining play and playing which were brought together in his book *Playing and Reality*. Winnicott, D.W. (1958) The capacity to be alone, in *The Maturational Processes and the Facilitating Environment*. London: Hogarth Press; Winnicott, D.W. (1971) *Playing and Reality*. London: Routledge, especially Chapters 3 to 6.

[3] Individuated does not mean individualised. Even in traditional societies the capacity to distinguish between self and other, me and not me, was a prerequisite for any kind of functioning.

[4] In an earlier chapter I referred to this as 'reagency'. Psychoanalysis has always insisted upon the fact of psychic bisexuality, that irrespective of our sex we all contain both male and female elements.

[5] Winnicott, Playing and Reality, see note 2 above, p. 85.

[6] I think this is a great phrase, to my knowledge one first used by John Rowan as the title for his pioneering book on Humanistic Psychology first published in 1976. Here I am using it in a related but slightly different sense to how John used it. Rowan, J. (1976) *Ordinary Ecstasy: Humanistic Psychology in Action*. London: Routledge & Kegan.

[7] MacCaig, N. (2009 [1964]) An Ordinary Day. In *The Poems of Norman MacCaig*. Edinburgh: Polygon.

[8] Sean O'Hagan interview with George Shaw, The Observer, Sunday 13 Feb, 2011. Available at https://www.theguardian.com/artanddesign/2011/feb/13/george-shaw-tile-hill-baltic-interview (accessed 11 September 2023).

[9] Jonathan Jones (2019) George Shaw review – the only artist who can unite England. *The Guardian*, 7 February, 2019. Available at: https://www.theguardian.com/artanddesign/2019/feb/07/george-shaw-a-corner-of-a-foreign-field-review-england-brexit-holburne-bath (accessed 11 September 2023).

[10] Bishop, J. and Hoggett, P. (1986) *Organising Around Enthusiasms*. London: Comedia, p. 128.

EXERCUSION

MEDITATIONS

When I was 18 and lived in Letchworth, a small town in Hertfordshire, I'd travel down the A1 every month to meet what in those days (the late 60's) was referred to as a guru. Afterwards I would endlessly repeat the mantra that he gave me in my search for enlightenment. Progress seemed slow, but one day I noticed in a music mag a review of an album called Meditations by someone called John Coltrane. Maybe this would help I thought, so I sent off for the album.

To say that the music was not what I expected would be an understatement. Far from something peaceful and calming the sounds coming from my record player were so extraordinary that I remember my father examining the stylus to check that it wasn't damaged. He then concluded that the musicians were just taking the piss. Reluctantly I tended to agree with him and put the album away, bringing it out occasionally to entertain a mate when we were drunk or to impress upon a girlfriend the uniqueness of my musical sensibility.

Oddly enough in the years that followed I found it growing on me. I'd suddenly find myself remembering a phrase or passage and before long I realised that it was haunting me.

Meditations is an ugly, fierce and discomforting piece of music and, I have to say, for me it is the most extraordinary thing of beauty as well. Six black men went into a studio and without rehearsing and in a single take made music like it had never been made before. To improvise like this they had to listen to each other in a different way, they had to let the sounds come and take them and shape them and move them. John Coltrane, Rashied Ali, Jimmy Garrison, Pharoah Sanders, McCoy Tyner and Elvin Jones. I have repeated their names countless times, my own handcrafted mantra.

If you want to know what Black Power sounded like in its early days, then this is it. Not just an angry political fierceness but an immense spiritual power – turbulent, unsettled and unsettling. Not the peace and love I was seeking back then but something much more real.

For more than 50 years Meditations has continued to haunt me. To give an example, just occasionally I have the discipline to get up at about 4.00 am in the early summer and listen to the dawn chorus in our valley. It's May and there's a mistle thrush soloing and I can hear Coltrane as the bird goes through its repertoire of riffs and verses against the background cacophony of chaffinches, tits, blackcaps and dunnocks.

Just knowing that this music exists has given me great solace and I'm sure has helped me be steadfast, to keep on going on in times such as the 1980's which, in the UK, were as bleak as today. In response to the 1980s I wrote Partisans in an Uncertain World *in 1992 in which I explored the challenge of remaining ethically and politically engaged in the face of what could seem like overwhelming odds. Three decades later and I have found myself writing similar book, this time in relation to the climate crisis. I recommend listening to* Meditations, *it is the one disk I would save on my desert island as the sea level starts to rise.*